T0142948

Wireless Networks

Series editor

Xuemin (Sherman) Shen
University of Waterloo, Waterloo, Ontario, Canada

More information about this series at http://www.springer.com/series/14180

Min Chen • Shigang Chen

RFID Technologies
for Internet of Things

 Springer

Min Chen
Department of Computer and Information
University of Florida
Gainesville, FL, USA

Shigang Chen
Department of Computer and
 Information Science
University of Florida
Gainesville, FL, USA

This work is supported in part by the National Science Foundation under grants CNS-1409797 and STC-1562485.

ISSN 2366-1186   ISSN 2366-1445   (electronic)
Wireless Networks
ISBN 978-3-319-83718-5   ISBN 978-3-319-47355-0   (eBook)
DOI 10.1007/978-3-319-47355-0

Printed on acid-free paper

This Springer imprint is published by Springer Nature
The registered company is Springer International Publishing AG
The registered company address is: Gewerbestrasse 11, 6330 Cham, Switzerland

Contents

Chapter 1
Introduction

1.1 Internet of Things

Internet of Things (IoT) [26] is a new networking paradigm for cyber-physical systems that allow physical objects to collect and exchange data. In the IoT, physical objects and cyber-agents can be sensed and controlled remotely across existing network infrastructure, which enables the integration between the physical world and computer-based systems and therefore extends the Internet into the real world. IoT can find numerous applications in smart housing, environmental monitoring, medical and health care systems, agriculture, transportation, etc. Because of its significant application potential, IoT has attracted a lot of attention from both academic research and industrial development.

1.2 RFID Technologies

Generally, every physical object in the IoT needs to be augmented with some auto-ID technologies such that the object can be uniquely identified. Radio Frequency Identification (RFID) [12] is one of the most widely used auto-ID technologies. RFID technologies integrate simple communication, storage, and computation components in attachable tags that can communicate with readers wirelessly over a distance. Therefore, RFID technologies provide a simple and cheap way of connecting physical objects to the IoT—as long as an object carries a tag, it can be identified and tracked by readers.

RFID technologies have been pervasively used in numerous applications, such as inventory management, supply chain, product tracking, transportation, logistics, and toll collection [1, 3, 8, 10, 16–19, 21–24, 27, 29, 32, 35, 37–39]. According to a market research conducted by IDTechEx [30], the market size of RFID has reached $8.89 billion in 2014, and is projected to rise to $27.31 billion after a decade.

© Springer International Publishing AG 2016
M. Chen, S. Chen, *RFID Technologies for Internet of Things*,
Wireless Networks, DOI 10.1007/978-3-319-47355-0_1

Typically, an RFID system consists of a large number of RFID tags, one or multiple RFID readers, and a backend server. Today's commercial tags can be classified into three categories: (1) passive tags, which are powered by the radio wave from an RFID reader and communicate with the reader through backscattering; (2) active tags, which are powered by their own energy sources; and (3) semi-active tags, which use internal energy sources to power their circuits while communicating with the reader through backscattering. As specified in EPC Class-1 Gen-2 (C1G2) protocol [12], each tag has a unique ID identifying the object it is attached to. The object can be a vehicle, a product in a warehouse, an e-passport that carries personal information, a medical device that records a patient's health data, or any other physical object in IoT. The integrated transceiver of each tag enables it to transmit and receive radio signals. Therefore, a reader can communicate with a tag over a distance as long as the tag is located in its interrogation area. However, communications amongst RFID tags are generally not feasible due to their low transmission power. The emerging networked tags [13, 14] bring a fundamental enhancement to RFID tags by enabling tags to communicate with each other. The networked tags are integrated with energy-harvesting components that can harvest energy from surrounding environment.

The widespread use of RFID tags in IoT brings about new issues on efficiency, security, and privacy that are quite different from those in traditional networking systems [7, 36]. This book presents several state-of-the-art RFID protocols that aim at improving the efficiency, security, and privacy of the IoT.

1.3 Tag Search Problem

Given a set of IDs for the wanted tags, the tag search problem is to identify which wanted tags are existing in an RFID system [9, 11]. Note that there may exist other tags that do not belong to the set. As an example, a manufacturer finds that some of its products, which have been distributed to different warehouses, may be defective, and wants to recall them for further inspection. Since each product in the IoT carries a tag, and the manufacturer knows all tag IDs of those defective products, it can perform tag search in each warehouse to identify the products that need to be recalled.

To meet the stringent delay requirements of real-world applications, time efficiency is a critical performance metric for the RFID tag search problem. For example, it is highly desirable to make the search quick in a busy warehouse as lengthy searching process may interfere with other activities that move things in and out of the warehouse. The only prior work studying this problem is called CATS [40], which, however, does not work well under some common conditions (e.g., if the size of the wanted set is much larger than the number of tags in the coverage area of the reader).

We present a fast tag search method based on a new technique called filtering vectors. A filtering vector is a compact one-dimension bit array constructed from tag IDs, which can be used for filtering unwanted tags. Using the filtering vectors, we design, analyze, and evaluate a novel iterative tag search protocol, which progressively improves the accuracy of search result and reduces the time of each iteration to a minimum by using the information learned from previous iterations. Given an accuracy requirement, the iterative protocol will terminate after the search result meets the accuracy requirement. We show that our protocol performs much better than the CATS protocol and other alternatives used for comparison. In addition, our protocol can be extended to work under noisy channel with a modest increase in execution time.

1.4 Anonymous RFID Authentication

The proliferation of RFID tags in their traditional ways makes their carriers trackable. Should future tags penetrate into everyday products in the IoT and be carried around (oftentimes unknowingly), people's privacy would become a serious concern. A typical tag will automatically transmit its ID in response to the query from a nearby reader. If we carry tags in our pockets or by our cars, these tags will give off their IDs to any readers that query them, allowing others to track us. As an example, for a person whose car carries a tag (automatic toll payment [35] or tagged plate [34]), he may be unknowingly tracked over years by toll booths or others who install readers at locations of interest to learn when and where he has been. To protect the privacy of tag carriers, we need to invent ways of keeping the usefulness of tags while doing so anonymously.

Many RFID applications such as toll payment require authentication. A reader will accept a tag's information only after authenticating the tag and vice versa. Anonymous authentication should prohibit the transmission of any identifying information, such as tag ID, key identifier, or any fixed number that may be used for identification purpose. As a result, there comes the challenge that how can a legitimate reader efficiently identify the right key for authentication without any identifying information of the tag?

The importance and challenge of anonymous authentication attract much attention from the RFID research community. Many anonymous authentication protocols have been designed. However, we will show that all prior work has some potential problems, either incurring high computation or communication overhead, or having security or functional concern. Moreover, most prior work, if not all, employs cryptographic hash functions, which requires considerable hardware [2], to randomize authentication data in order to make the tags untrackable. The high hardware requirement makes them not suited for low-cost tags with limited hardware resource. Hence, designing anonymous authentication protocols for low-cost tags remains an open and challenging problem [4].

In our design, we make a fundamental shift from the traditional paradigm for anonymous RFID authentication [5]. First, we release the resource-constrained RFID tags from implementing any complicated functions (e.g., cryptographic hashes). Since the readers are not needed in a large quantity as tags do, they can have much more hardware resource. Therefore, we follow the asymmetry design principle to push most complexity to the readers while leaving the tags as simple as possible. Second, we develop a novel technique to generate random tokens on demand for anonymous authentication. Our protocol only requires $O(1)$ communication overhead and online computation overhead per authentication for both readers and tags, which is a significant improvement over the prior art. Hence, our protocol is scalable to large RFID systems. Finally, extensive theoretic analysis, security analysis, simulations, and statistical randomness tests are provided to verify the effectiveness of our protocol.

1.5   Identification of Networked Tags

The emerging networked tags promise to bring a fundamental enhancement to traditional RFID tags by enabling tags to communicate with each other. An example of such tags is a new class of ultra-low-power energy-harvesting networked tags designed and prototyped at Columbia University [13, 14]. Tagged objects that are not traditionally networked among themselves, e.g., warehouse products, books, furniture, and clothing, can now form a network [15], which provides great flexibility in applications. Consider a large warehouse where a great number of readers and antennas must be deployed to provide full coverage. Such a system deployment can be very costly. Moreover, obstacles and piles of tagged objects may prevent signals from penetrating into every corner of the deployment, causing a reader to fail in accessing some of the tags. This problem will be solved if the tags can relay transmissions towards the otherwise-inaccessible reader. Hence, we envision that networked tags will play an important role in the IoT as an enhancement to the current RFID technologies.

The new feature of networked tags opens up new research opportunities. We focus on the tag identification problem, which is to collect the IDs of all tags in a system [6]. This is the most fundamental problem for RFID systems, but has not been studied in the context of networked tags, where one can take advantage of the networking capability to facilitate the ID collection process, e.g., collecting IDs of tags that are not within the reader's coverage. Beyond the coverage of the readers, the networked tags are powered by batteries or rechargeable energy sources that opportunistically harvest solar, piezoelectric, or thermal energy from surrounding environment [14, 20]. Therefore, energy efficiency is a first-order performance criterion for operations carried out by networked tags. In addition, we should also make the process of tag identification time-efficient so that it can scale to a large tag system where the communication channel works at a very low rate for energy conservation.

The existing RFID identification [25, 28, 31, 33] protocols cannot be applied to identifying networked tags because they assume that the readers can reach all tags directly. We make a fundamental shift in the protocol design for networked tags, and present two solutions: a contention-based ID collection protocol and a serialized

ID collection protocol. We reveal that the traditional contention-based protocol design will incur too much overhead in multihop networked tag systems due to progressively increased collision in the network towards a reader, which results in excessive energy cost. In contrast, a reader-coordinated design that attempts to serialize tag transmissions performs much better. Moreover, we show that load balancing is critical in controlling the worst-case energy cost incurring to the tags. We find that the worst-case energy cost is high for both contention-based and serialized protocol designs due to imbalanced load in the network. For the serialized ID collection protocol, however, we are able to provide a solution based on serial numbers that balance the load and reduce the worst-case energy cost.

1.6   Outline of the Book

The rest of the book is organized as follows. Chapter 2 presents an efficient tag search protocol based on filtering vectors and evaluates the impact of channel noise on its performance. Chapter 3 introduces the problem of anonymous authentication in RFID systems. We present a lightweight anonymous authentication protocol using dynamically generated tokens. The protocol only requires constant communication and computation overhead for both readers and tags. Chapter 4 discusses the problem of identifying networked tags. Two tag identification protocols are given and compared in detail.

References

1. AEI Technology. Available at http://www.aeitag.com/aeirfidtec.html (2008)
2. Bogdanov, A., Leander, G., Paar, C., Poschmann, A., Robshaw, M.J.B., Seurin, Y.: Hash functions and RFID tags: mind the gap. In: Proceedings of CHES, pp. 283–299 (2008)
3. Bu, K., Xiao, B., Xiao, Q., Chen, S.: Efficient pinpointing of misplaced tags in large RFID systems. In: Proceedings of IEEE SECON, pp. 287–295 (2011)
4. Chen, M., Chen, S.: An efficient anonymous authentication protocol for RFID systems using dynamic tokens. In: Proceedings of IEEE ICDCS (2015)
5. Chen, M., Chen, S.: ETAP: enable lightweight anonymous RFID authentication with O(1) overhead. In: Proceedings of IEEE ICNP (2015)
6. Chen, M., Chen, S.: Identifying state-free networked tags. In: Proceedings of IEEE ICNP (2015)
7. Chen, S., Deng, Y., Attie, P., Sun, W.: Optimal deadlock detection in distributed systems based on locally constructed wait-for graphs. In: Proceedings of IEEE INFOCOM, pp. 613–619 (1996)
8. Chen, S., Zhang, M., Xiao, B.: Efficient information collection protocols for sensor-augmented RFID networks. In: Proceedings of IEEE INFOCOM, pp. 3101–3109 (2011)
9. Chen, M., Luo, W., Mo, Z., Chen, S., Fang, Y.: An efficient tag search protocol in large-scale RFID systems. In: Proceedings of IEEE INFOCOM, pp. 1325–1333 (2013)
10. Chen, M., Chen, S., Xiao, Q.: Pandaka: a lightweight cipher for RFID systems. In: Proceedings of IEEE INFOCOM, pp. 172–180 (2014)

11. Chen, M., Luo, W., Mo, Z., Chen, S., Fang, Y.: An efficient tag search protocol in large-scale RFID systems with noisy channel. IEEE/ACM Trans. Networking **PP**(99), 1–1 (2015)
12. EPC Radio-Frequency Identity Protocols Class-1 Gen-2 UHF RFID Protocol for Communications at 860MHz-960MHz, EPCglobal. Available at http://www.epcglobalinc.org/uhfclg2 (2011)
13. Gorlatova, M., Kinget, P., Kymissis, I., Rubenstein, D., Wang, X., Zussman, G.: Challenge: ultra-low-power energy-harvesting active networked tags (EnHANTs). In: Proceedings of ACM Mobicom, pp. 253–260 (2009)
14. Gorlatova, M., Margolies, R., Sarik, J., Stanje, G., Zhu, J., Vigraham, B., Szczodrak, M., Carloni, L., Kinget, P., Kymissis, I., Zussman, G.: Prototyping energy harvesting active networked tags (EnHANTs). In: Proceedings of IEEE INFOCOM mini-conference (2013)
15. Kinget, P., Kymissis, I., Rubenstein, D., Wang, X., Zussman, G.: Energy harvesting active networked tags (EnHANTs) for ubiquitous object networking. IEEE Trans. Wirel. Commun. **17**(6), 18–25 (2010)
16. Lee, C.H., Chung, C.W.: Efficient storage scheme and query processing for supply chain management using RFID. In: Proceedings of ACM SIGMOD (2008)
17. Li, Y., Ding, X.: Protecting RFID communications in supply chains. In: Proceedings of IEEE ASIACCS (2007)
18. Li, T., Chen, S., Ling, Y.: Efficient protocols for identifying the missing tags in a large RFID system. IEEE/ACM Trans. Networking **21**(6), 1974–1987 (2013)
19. Liu, J., Xiao, B., Bu, K., Chen, L.: Efficient distributed query processing in large RFID-enabled supply chains. In: Proceedings of IEEE INFOCOM, pp. 163–171 (2013)
20. Liu, V., Parks, A., Talla, V., Gollakota, S., Wetherall, D., Smith, J.R.: Ambient backscatter: wireless communication out of thin air. In: Proceedings of ACM SIGCOMM, pp. 39–50 (2013)
21. Liu, J., Chen, M., Xiao, B., Zhu, F., Chen, S., Chen, L.: Efficient RFID grouping protocols. IEEE/ACM Trans. Networking **PP**(99), 1–1 (2016)
22. Luo, W., Chen, S., Li, T.: Probabilistic missing-tag detection and energy-time tradeoff in large-scale RFID systems. In: Proceedings of ACM Mobihoc (2012)
23. Luo, W., Qiao, Y., Chen, S.: An efficient protocol for RFID multigroup threshold-based classification. In: Proceedings of IEEE INFOCOM, pp. 890–898 (2013)
24. Luo, W., Qiao, Y., Chen, S., Chen, M.: An efficient protocol for RFID multigroup threshold-based classification based on sampling and logical bitmap. IEEE/ACM Trans. Networking **24**(1), 397–407 (2016)
25. Myung, J., Lee, W.: Adaptive splitting protocols for RFID tag collision arbitration. In: Proceedings of ACM Mobihoc (2006)
26. Network Everything. Available at http://openinterconnect.org
27. Ni, L., Liu, Y., Lau, Y.C.: Landmarc: indoor location sensing using active RFID. In: Proceedings of IEEE PerCom (2003)
28. Qian, C., Liu, Y., Ngan, H., Ni, L.M.: ASAP: scalable identification and counting for contactless RFID systems. In: Proceedings of IEEE ICDCS (2010)
29. Qiao, Y., Chen, S., Li, T.: Energy-efficient polling protocols in RFID systems. In: Proceedings of ACM Mobihoc (2011)
30. RFID Report. Available at http://www.idtechex.com/research/reports/rfid-forecasts-players-and-opportunities-2014-2024-000368.asp (2013)
31. Shahzad, M., Liu, A.X.: Probabilistic optimal tree hopping for RFID identification. In: Proceedings of ACM SIGMETRICS, pp. 293–304 (2013)
32. Sheng, B., Tan, C., Li, Q., Mao, W.: Finding popular categories for RFID tags. In: Proceedings of ACM Mobihoc (2008)
33. Sheng, B., Li, Q., Mao, W.: Efficient continuous scanning in RFID systems. In: Proceedings of IEEE INFOCOM (2010)
34. SINIAV. Available at http://roadpricing.blogspot.com/2012/08/brazil-to-have-compulsory-toll-tags-by.html (2012)
35. Sun Pass. Available at https://www.sunpass.com/index

36. Xia, Y., Chen, S., Cho, C., Korgaonkar, V.: Algorithms and performance of load-balancing with multiple hash functions in massive content distribution. Comput. Netw. **53**(1), 110–125 (2009)
37. Xiao, Q., Chen, M., Chen, S., Zhou, Y.: Temporally or spatially dispersed joint RFID estimation using snapshots of variable lengths. In: Proceedings of ACM Mobihoc (2015)
38. Xiao, Q., Chen, S., Chen, M.: Joint property estimation for multiple RFID tag sets using snapshots of variable lengths. In: Proceedings of ACM Mobihoc (2016)
39. Zhang, Z., Chen, S., Ling, Y., Chow, R.: Capacity-aware multicast algorithms on heterogeneous overlay networks. IEEE Trans. Parallel Distrib. Syst. **17**(2), 135–147 (2006)
40. Zheng, Y., Li, M.: Fast tag searching protocol for large-scale RFID systems. IEEE/ACM Trans. Networking **21**(3), 924–934 (2012)

Chapter 2
Efficient Tag Search in Large RFID Systems

This chapter introduces the tag search problem in large RFID systems. A new technique called filtering vector is designed to reduce the transmission overhead during search process, thereby improving the time efficiency. Based on this technique, we present an iterative tag search protocol. Some tags are filtered out in each round and the search process will eventually terminate when the result meets a given accuracy requirement. Moreover, the protocol is extended to work under noisy channel. The simulation results demonstrate that our protocol performs much better than the best existing work.

The rest of this chapter is organized as follows. Section 2.1 gives the system model and the problem statement. Section 2.2 briefly introduces the related work. Section 2.3 describes our new protocol in detail. Section 2.4 addresses noisy wireless channel. Section 2.5 evaluates the performance of our protocol by simulations. Section 2.6 gives the summary.

2.1 System Model and Problem Statement

2.1.1 System Model

We consider an RFID system consisting of one or more readers, a backend server, and a large number of tags. Each tag has a unique 96-bit ID according to the EPC global Class-1 Gen-2 (C1G2) standard [9]. A tag is able to communicate with the reader wirelessly and perform some computations such as hashing. The backend server is responsible for data storage, information processing, and coordination. It is capable of carrying out high-performance computations. Each reader is connected to the backend server via a high speed wired or wireless link. If there are many readers (or antennas), we divide them into non-interfering groups and any RFID protocol can be performed for one group at a time, with the readers in that group

© Springer International Publishing AG 2016
M. Chen, S. Chen, *RFID Technologies for Internet of Things*,
Wireless Networks, DOI 10.1007/978-3-319-47355-0_2

executing the protocol in parallel. The readers in each group can be regarded as an integrated unit, still called a reader for simplicity. Many works regarding multi-reader coordination can be found in literature [5, 7, 17].

In practice, the tag-to-reader transmission rate and the reader-to-tag transmission rate may be different and subject to the environment. For example, as specified in the EPC global Class-1 Gen-2 standard, the tag-to-reader transmission rate is 40–640kbps in the FM0 encoding format or 5–320kbps in the Miller modulated subcarrier encoding format, while the reader-to-tag transmission rate is about 26.7–128kbps. However, to simplify our discussions, we assume the tag-to-reader transmission rate and the reader-to-tag transmission rate are the same, and it is straightforward to adapt our protocol for asymmetric transmission rates.

2.1.2 Time Slots

The RFID reader and the tags in its coverage area use a framed slotted MAC protocol to communicate. We assume that clocks of the reader and all tags in the RFID system are synchronized by the reader's signal. During each frame, the communication is initialized by the reader in a request-and-response mode, namely the reader broadcasts a request with some parameters to the tags and then waits for the tags to reply in the subsequent time slots.

Consider an arbitrary time slot. We call it an empty slot if no tag replies in this slot, or a busy slot if one or more tags respond in this slot. Generally, a tag just needs to send one-bit information to make the channel busy such that the reader can sense its existence. The reader uses "0" to represent an empty slot with an idle channel and "1" for a busy slot with a busy channel. The length of a slot for a tag to transmit a one-bit short response is denoted as t_s. Note that t_s can be set larger than the time of one-bit data transmission for better tolerance of clock drift in tags. Some prior RFID work needs another type of slots for transmission of tag IDs, which will be introduced shortly.

2.1.3 Problem Statement

Suppose we are interested in a known set of tag IDs $X = \{x_1, x_2, x_3, \cdots\}$, each $x_i \in X$ is called a *wanted tag*. For example, the set may contain tag IDs on a certain type of products under recall by a manufacturer. Let $Y = \{y_1, y_2, y_3, \cdots\}$ be the set of tags within the coverage area of an RFID system (e.g., in a warehouse). Each x_i or y_i represents a tag ID. The tag search problem is to identify the subset W of wanted tags that are present in the coverage area. Namely, $W \subseteq X$. Since each tag in W is in the coverage area, $W \subseteq Y$. Therefore, $W = X \cap Y$. We define the intersection ratio of X and Y as

$$R_{INTS} = \frac{|W|}{\min\{|X|, |Y|\}}. \tag{2.1}$$

Exactly finding W can be expensive if X and Y are very large. It is much more efficient to find W approximately, allowing small bounded error [28]—all wanted tags in the coverage area must be identified, but a few wanted ones that are not in the coverage may be accidentally included.[1]

Our solution performs iteratively. Each round rules out some tags in X when it becomes certain that they are not in the coverage area (i.e., Y), and it also rules out some tags in Y when it becomes certain that they are not wanted ones in X. These ruled-out tags are called non-candidate tags. Other tags that remain possible to be in both X and Y are called candidate tags. At the beginning, the search result is initialized to all wanted tags X. As our solution is iteratively executed, the search result shrinks towards W when more and more non-candidates are ruled out.

Let W^* be the final search result. We have the following two requirements:

1. All wanted tags in the coverage area must be detected, namely $W \subseteq W^*$.
2. A false positive occurs when a tag in $X - W$ is included in W^*, i.e., a tag not in the coverage area is kept in the search result by the reader.[2] The false-positive ratio is the probability for any tag in $X - W$ to be in W^* after the execution of a search protocol. We want to bound the false-positive ratio by a pre-specified system requirement P_{REQ}, whose value is set by the user. In other words, we expect

$$\frac{|W^* - W|}{|X - W|} \leq P_{REQ}. \tag{2.2}$$

Notations used in this chapter are given in Table 2.1 for quick reference.

2.2 Related Work

2.2.1 Tag Identification

A straightforward solution for the tag search problem is identifying all existing tags in Y. After that, we can apply an intersection operation $X \cap Y$ to compute W. EPC C1G2 standard assumes that the reader can only read one tag ID at a time. Dynamic Framed Slotted ALOHA (DFSA) [4, 8, 19–21] is implemented to deal with tag collisions, where each frame consists of a certain number of equal-duration slots.

---

[1]If perfect accuracy is necessary, a post step may be taken by the reader to broadcast the identified IDs. As the wanted tags in the coverage reply after hearing their IDs, those mistakenly included tags can be excluded due to non-response to these IDs.

[2]The nature of our protocol guarantees that all tags in $Y - W$ are not included in W^*.

Table 2.1 Notations

Symbols	Descriptions
X	Set of wanted tags
Y	Set of tags in the RFID system
W	Intersection of X and Y, i.e., $W = X \cap Y$
X_i	Set of remaining candidate tags in X, i.e., search result
	at the beginning of the ith round of our protocol;
Y_i	Set of remaining candidate tags in Y at the beginning
	of the ith round of our protocol
U_i	Difference between X_i and W, i.e., $U_i = X_i - W$
V_i	Difference between Y_i and W, i.e., $V_i = Y_i - W$
$|\cdot|$	Cardinality of the set
$h(\cdot)$	A uniform hash function
$FV(\cdot)$	Filtering vector of a set

It is proved that the theoretical upper bound of identification throughput using DFSA is approximately $\frac{1}{e}$ tags per slot (e is the natural constant), which is achieved when the frame size is set equal to the number of unidentified tags [25]. As specified in EPC C1G2, each slot consists of the transmissions of a QueryAdjust or QueryRep command from the reader, one tag ID, and two 16-bit random numbers: one for the channel reservation (collision avoidance) sent by the tags, and the other for ACK/NAK transmitted by the reader. We denote the duration of each slot for tag identification as t_l. Therefore, the lower bound of identification time for tags in Y using DFSA is

$$T_{DFSA} = e \times |Y| \times t_l. \tag{2.3}$$

One limitation of the current DFSA is that the information contained in collision slots is wasted. Some recent work [3, 12, 15, 16, 24, 27] focuses on Collision Recovery (CR) techniques, which enable the resolution of multiple tag IDs from a collision slot. Benefiting from the CR techniques, the identification throughput can be dramatically improved up to 3.1 tags per slot in [16]. Suppose the throughput is υ tags per slot after adopting the CR techniques. The lower bound for identification time is

$$T_{CR} = \frac{|Y|}{\upsilon} \times t_l. \tag{2.4}$$

Note that after employing the CR techniques the real duration of each slot can be longer than t_l. The reason is that the reader may need to acknowledge multiple tags and the tags may need to send extra messages to facilitate collision recovery.

2.2.2 Polling Protocol

The polling protocol provides an alternative solution to the tag search problem. Instead of collecting all IDs in Y, the reader can broadcast the IDs in X one by one. Upon receiving an ID, each tag checks whether the received ID is identical to its own. If so, the tag transmits a one-bit short response to notify the reader about its presence; otherwise, the tag keeps silent. Hence, the execution time of the polling protocol is

$$T_{Polling} = |X| \times (t_{id} + t_s), \tag{2.5}$$

where t_{id} is the time cost for the reader to broadcast a tag ID.

The polling protocol is very efficient when $|X|$ is small. However, it also has serious limitations. First, it does not work well when $|X| \gg |Y|$. Second, the energy consumption of tags (particularly when active tags are used) is significant because tags in Y have to continuously listen to the channel and receive a large number of IDs until its own ID is received.

2.2.3 CATS Protocol

To address the problems of the tag identification and polling protocols, Zheng et al. design a two-phase protocol named *Compact Approximator based Tag Searching protocol* (CATS) [28], which is the most efficient solution for the tag search problem to date.

The main idea of the CATS protocol is to encode tag IDs into a Bloom filter and then transmit the Bloom filter instead of the IDs. In its first phase, the reader encodes all IDs of wanted tags in X into an L_1-bit Bloom filter, and then broadcasts this filter together with some parameters to tags in the coverage area. Having received this Bloom filter, each tag tests whether it belongs to the set X. If the answer is negative, the tag is a non-candidate and will keep silent for the remaining time. After the filtration of phase one, the number of candidate tags in Y is reduced. During the second phase, the remaining candidate tags in Y report their presence in a second L_2-bit Bloom filter constructed from a frame of time slots t_s. Each candidate tag transmits in k slots that it is mapped to. Listening to channel, the reader builds the Bloom filter based on the status of the time slots: "0" for an idle slot where no tag transmits, and "1" for a busy slot where at least one tag transmits. Using this Bloom filter, the reader conducts filtration for the IDs in X to see which of them belong to Y, and the result is regarded as $X \cap Y$.

With a pre-specified false-positive ratio requirement P_{REQ}, the CATS protocol uses the following optimal settings for L_1 and L_2:

$$L_1 = |X| \log_\phi \left(-\frac{\alpha |X|}{\beta |Y| \ln P_{REQ}} \right), \tag{2.6}$$

$$L_2 = \frac{|X|}{\ln \phi} \left(\ln P_{REQ} - \frac{\alpha}{\beta} \right), \tag{2.7}$$

where ϕ is a constant that equals 0.6185, α and β are constants pertaining to the reader-to-tag transmission rate and the tag-to-reader transmission rate, respectively. In CATS, the authors assume t_s is the time needed to delivering one-bit data, and $\alpha = \beta$, i.e., the reader-to-tag transmission rate and the tag-to-reader transmission rate are identical. Therefore, the total search time of the CATS protocol is

$$\begin{aligned}
T_{CATS} &= (L_1 + L_2) \times t_s \\
&= |X| \left(\log_\phi \left(\frac{-|X|}{|Y| \ln P_{REQ}} \right) + \frac{\ln P_{REQ} - 1}{\ln \phi} \right) \times t_s.
\end{aligned} \tag{2.8}$$

2.3 A Fast Tag Search Protocol Based on Filtering Vectors

This section presents an Iterative Tag Search Protocol (ITSP) to solve the tag search problem in large-scale RFID systems. We will ignore channel error for now and delay this subject to Sect. 2.4.

2.3.1 Motivation

Although the CATS protocol takes a significant step forward in solving the tag search problem, it still has several important drawbacks. First, when optimizing the Bloom filter sizes L_1 and L_2, CATS approximates $|X \cap Y|$ simply as $|X|$. This rough approximation may cause considerable overhead when $|X \cap Y|$ deviates significantly from $|X|$.

Second, it assumes that $|X| < |Y|$ in its design and formula derivation. In reality, the number of wanted tags may be far greater than the number in the coverage area of an RFID system. For example, there may be a huge number $|X|$ of tagged products that are under recall, but as the products are distributed to many warehouses, the number $|Y|$ of tags in a particular warehouse may be much smaller than $|X|$. Although CATS can still work under conditions of $|X| \gg |Y|$, it will become less efficient as our simulations will demonstrate.

Third, the performance of CATS is sensitive to the false-positive ratio requirement P_{REQ}. The performance deteriorates when the value of P_{REQ} is very small. While the simulations in [28] set $P_{REQ} = 5\%$, its value may have to be much smaller in some practical cases. For example, suppose $|X| = 100,000$, and $|W| = 1000$. If we set $P_{REQ} = 5\%$, the number of wanted tags that are falsely claimed to be in Y by CATS will be up to $|X - W| \times P_{REQ} = 4995$, far more than the 1000 wanted tags that are actually in Y.

We will show that an iterative way of implementing Bloom filters is much more efficient than the classical way that the CATS protocol adopts.

2.3.2   Bloom Filter

A Bloom filter is a compact data structure that encodes the membership for a set of items. To represent a set $S = \{e_1, e_2, \cdots, e_n\}$ using a Bloom filter, we need a bit array of length l in which all bits are initialized to zeros. To encode each element $e \in S$, we use k hash functions, $h_1, h_2, \cdots, h_k$, to map the element randomly to k bits in the bit array, and set those bits to ones. For membership lookup of an element b, we again map the element to k bits in the array and see if all of them are ones. If so, we claim that b belongs to S; otherwise, it must be true that $b \notin S$. A Bloom filter may cause false positive: a non-member element is falsely claimed as a member in S. The probability for a false positive to occur in a membership lookup is given as follows [2, 23]:

$$P_B = \left(1 - \left(1 - \frac{1}{l}\right)^{kn}\right)^k \approx \left(1 - e^{-kn/l}\right)^k. \qquad (2.9)$$

When $k = \ln 2 \times \frac{l}{n}$, P_B is approximately minimized to $\left(\frac{1}{2}\right)^k = \left(\frac{1}{2}\right)^{\ln 2 \frac{l}{n}}$. In order to achieve a target value of P_B, the minimum size of the filter is $-\frac{\ln P_B}{(\ln 2)^2} n$.

CATS sends one Bloom filter from the reader to tags and another Bloom filter from tags back to the reader. Consider the first Bloom filter that encodes X. As $n = |X|$, the filter size is $-\frac{\ln P_B}{(\ln 2)^2} |X|$. As an example, to achieve $P_B = 0.001$, the size becomes $14.4 \times |X|$ bits. Similarly, the size of the second filter from tags to the reader is also related to the target false-positive probability.

Below we show that the overall size of the Bloom filter can be significantly reduced by reconstructing it as filtering vectors and then iteratively applying these vectors.

2.3.3   Filtering Vectors

A Bloom filter can also be implemented in a segmented way. We divide its bit array into k equal segments, and the ith hash function will map each element to a random bit in the ith segment, for $i \in [1...k]$. We name each segment as a filtering vector (FV), which has l/k bits. The following formula gives the false-positive probability

of a single filtering vector, i.e., the probability for a non-member to be hashed to a
"1" bit in the vector:

$$P_{FV} = 1 - \left(1 - \frac{1}{l/k}\right)^n \approx 1 - e^{-kn/l}. \tag{2.10}$$

Since there are k independent segments, the overall false-positive probability of a
segmented Bloom filter is

$$P_{FP} = (P_{FV})^k \approx \left(1 - e^{-kn/l}\right)^k, \tag{2.11}$$

which is approximately the same as the result in (2.9). It means that the two ways
of implementing a Bloom filter have similar performance. The value P_{FP} is also
minimized when $k = \ln 2 \times \frac{l}{n}$. Hence, the optimal size of each filtering vector is

$$\frac{l}{k} = \frac{n}{\ln 2}, \tag{2.12}$$

which results in

$$P_{FV} \approx \frac{1}{2}. \tag{2.13}$$

Namely, each filtering vector on average filters out half of non-members.

Figure 2.1 illustrates the concept of filtering vectors. Suppose we have two
elements a and b, two hash function h_1 and h_2, and an 8-bit bit array. First,
suppose $h_1(a) \bmod 8 = 1$, $h_1(b) \bmod 8 = 7$, $h_2(a) \bmod 8 = 5$, $h_2(b) \bmod 8 = 2$,
and we construct a Bloom filter for a and b in the upper half of the figure. Next, we
divide the bit array into two 4-bit filtering vectors, and apply h_1 to the first segment
and h_2 to the second segment. Since $h_1(a) \bmod 4 = 1$, $h_1(b) \bmod 4 = 3$, $h_2(a) \bmod 4$
$= 1$, $h_2(b) \bmod 4 = 2$, we build the two filtering vectors in the lower half of the figure.

Fig. 2.1 Bloom filter and
filtering vectors

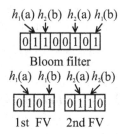

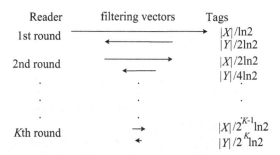

Fig. 2.2 Iterative use of filtering vectors. Each *arrow* represents one filtering vector, and the *length of the arrow* indicates the filtering vector's size, which is specified to the right. As the size shrinks in subsequent rounds, the total amount of data exchanged between the reader and the tags is significantly reduced

2.3.4   Iterative Use of Filtering Vectors

In this work, we use filtering vectors in a novel iterative way: Bloom filters between the reader and tags are exchanged in rounds; one filtering vector is exchanged in each round, and the size of filtering vector is continuously reduced in subsequent rounds, such that the overall size of each Bloom filter is much reduced.

Below we use a simplified example to explain the idea, which is illustrated in Fig. 2.2: Suppose there is no wanted tag in the coverage area of an RFID reader, namely $X \cap Y = \emptyset$. In round one, we firstly encode X in a filtering vector of size $|X|/\ln 2$ through a hash function h_1, and broadcast the vector to filter tags in Y. Using the same hash function, each candidate tag in Y knows which bit in the vector it is mapped to, and it only needs to check the value of that bit. If the bit is zero, the tag becomes a non-candidate and will not participate in the protocol execution further. The filtering vector reduces the number of candidate tags in Y to about $|Y| \times P_{FV} \approx |Y|/2$. Then a filtering vector of size $|Y|/(2\ln 2)$ is sent from the remaining candidate tags in Y back to the reader in a way similar to [28]: Each candidate tag hashes its ID to a slot in a time frame and transmit one-bit response in that slot. By listening to the states of the slots in the time frame, the reader constructs the filtering vector, "1" for busy slots and "0" for empty slots. The reader uses this vector to filter non-candidate tags from X. After filtering, the number of candidate tags remaining in X is reduced to about $|X| \times P_{FV} \approx |X|/2$. Only the candidate tags in X need to be encoded in the next filtering vector, using a different hash function h_2. Hence, in the second round, the size of the filtering vector from the reader to tags is reduced by half to $|X|/(2\ln 2)$, and similarly the size of the filtering vector from tags to the reader is also reduced by half to $|Y|/(4\ln 2)$. Repeating the above process, it is easy to see that in the i_{th} round, the size of the filtering vector from the reader to tags is $|X|/(2^{i-1}\ln 2)$, and the size of the filtering vector from tags to the reader is $|Y|/(2^i \ln 2)$. After K rounds, the total size of all filtering vectors from the reader to tags is

$$\frac{1}{\ln 2} \sum_{i=1}^{K} \frac{|X|}{2^{i-1}} < \frac{2|X|}{\ln 2}, \tag{2.14}$$

where $\frac{2|X|}{\ln 2}$ is an upper bound, regardless of the number K of rounds (i.e., regardless of the requirement on the false-positive probability). It compares favorably to CATS whose filter size, $-\frac{\ln P_B}{(\ln 2)^2}|X|$, grows inversely in P_B, and reaches $14.4 \times |X|$ bits when $P_B = 0.001$ in our earlier example.

Similarly, the total size of all filtering vectors from tags to the reader is

$$\frac{1}{\ln 2} \sum_{i=1}^{K} \frac{|Y|}{2^{i}} < \frac{|Y|}{\ln 2}, \tag{2.15}$$

and $P_{FP} = (P_{FV})^K \approx \left(\frac{1}{2}\right)^K$. We can make P_{FP} as small as we like by increasing n, while the total transmission overhead never exceeds $\frac{1}{\ln 2}(2|X| + |Y|)$ bits. The strength of filtering vectors in bidirectional filtration lies in their ability to reduce the candidate sets during each round, thereby diminishing the sizes of filtering vectors in subsequent rounds and thus saving time. Its power of reducing subsequent filtering vectors is related to $|X - W|$ and $|Y - W|$. The more the numbers of tags outside of W, the more they will be filtered in each round, and the greater the effect of reduction.

2.3.5 Generalized Approach

Unlike the CATS protocol, our iterative approach divides the bidirectional filtration in tag search process into multiple rounds. Before the ith round, the set of candidate tags in X is denoted as X_i ($\subseteq X$), which is also called the search result after the $(i-1)$th round. The final search result is the set of remaining candidate tags in X after all rounds are completed. Before the ith round, the set of candidate tags in Y is denoted as Y_i ($\subseteq Y$). Initially, $X_1 = X$ and $Y_1 = Y$. We define $U_i = X_i - W$ and $V_i = Y_i - W$, which are the tags to be filtered out. Because W is always a subset of both X_i and Y_i, we have

$$|U_i| = |X_i| - |W|$$
$$|V_i| = |Y_i| - |W|. \tag{2.16}$$

Instead of exchanging a single filtering vector at a time, we generalize our iterative approach by allowing multiple filtering vectors to be sent consecutively. Each round consists of two phases. In phase one of the ith round, the RFID reader broadcasts a number m_i of filtering vectors, which shrink the set of remaining candidate tags in Y from Y_i to Y_{i+1}. In phase two of the ith round, one filtering

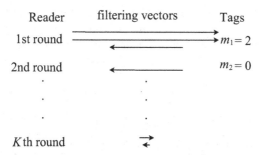

Fig. 2.3 Generalized approach. Each round has two phases. In phase one, the reader transmits zero, one, or multiple filtering vectors. In phase two, the tags send exactly one filtering vector to the reader. In the example shown by the figure, $m_1 = 2$ and $m_2 = 0$, which means there are two filtering vectors sent by the reader in the first round, while no filtering vector from the reader during the second round

vector is sent from the remaining candidate tags in Y_{i+1} back to the reader, which uses the received filtering vector to shrink its set of remaining candidates from X_i to X_{i+1}, setting the stage for the next round. This process continues until the false-positive ratio meets the requirement of P_{REQ}.

The values of m_i will be determined in the next subsection. If $m_i > 0$, multiple filtering vectors will be sent consecutively from the reader to tags in one round. If $m_i = 0$, no filtering vector is sent from the reader in this round. When this happens, it essentially allows multiple filtering vectors to be sent consecutively from tags to the reader (across multiple rounds). An illustration is given in Fig. 2.3.

2.3.6   Values of m_i

Let K be the total number of rounds. After all K rounds, we use X_{K+1} as our search result. There are in total K filtering vectors sent from tags to the reader. We know from Sect. 2.3.3 that each filtering vector can filter out half of non-members (in our case, tags in $X - W$). To meet the false-positive ratio requirement P_{REQ}, the following constraint should hold:

$$(P_{FV})^K \approx \left(\frac{1}{2}\right)^K \leq P_{REQ}. \tag{2.17}$$

Hence, the value of K is set to $\lceil -\frac{\ln P_{REQ}}{\ln 2} \rceil$. (We will discuss how to guarantee meeting the requirement P_{REQ} in Sect. 2.3.9.)

Next, we discuss how to set the values of m_i, $1 \leq i \leq K$, in order to minimize the execution time of each round. We use $FV(\cdot)$ to denote the filtering vector of a set. In phase one of the ith round, the reader builds m_i filtering vectors, denoted as

$FV_{i1}(X_i)$, $FV_{i2}(X_i)$, $\cdots$, $FV_{im_i}(X_i)$, which are consecutively broadcasted to the tags. From (2.12), we know the size of each filtering vector is $|X_i|/\ln 2$. After the filtration based on these vectors, the number of remaining candidate tags in Y_{i+1} is on average

$$
\begin{aligned}
|Y_{i+1}| &\approx |V_i| \times (P_{FV})^{m_i} + |W| \\
&\approx |V_i| \times (1/2)^{m_i} + |W| \\
&= |V_i|/2^{m_i} + |W|.
\end{aligned}
\tag{2.18}
$$

In phase two of the ith round, the tags in Y_{i+1} use a time frame of $\frac{1}{\ln 2} \times |Y_{i+1}|$ slots to report their presence. After receiving the responses, the reader builds a filtering vector, denoted as $FV_i(Y_{i+1})$. After the filtration based on $FV_i(Y_{i+1})$, the size of the search result X_{i+1} is on average

$$
\begin{aligned}
|X_{i+1}| &\approx |U_i| \times P_{FV} + |W| \\
&\approx |U_i|/2 + |W| \\
&= (|X_i| + |W|)/2.
\end{aligned}
\tag{2.19}
$$

We denote the transmission time of the ith round by $f(m_i)$. In order to make a fair comparison with CATS, we utilize the parameter setting that conforms with [28]. Therefore, $f(m_i) = \frac{1}{\ln 2} \times m_i \times |X_i| \times t_s + \frac{1}{\ln 2} \times |Y_{i+1}| \times t_s$, which is set to be:

$$
f(m_i) = \frac{t_s}{\ln 2} \left(m_i |X_i| + |V_i|/2^{m_i} + |W| \right).
\tag{2.20}
$$

To find the value of m_i that minimizes $f(m_i)$, we take the first-order derivative and set the right side to zero.

$$
\frac{df(m_i)}{dm_i} = \frac{t_s}{\ln 2} \left(|X_i| - \ln 2 |V_i|/2^{m_i} \right) = 0
\tag{2.21}
$$

Hence, the value of $f(m_i)$ is minimized when

$$
m_i = \frac{\ln(\ln 2 |V_i|/|X_i|)}{\ln 2}.
\tag{2.22}
$$

Because m_i cannot be a negative number, we reset $m_i = 0$ if $\frac{\ln(\ln 2 |V_i|/|X_i|)}{\ln 2} < 0$. Furthermore, m_i must be an integer. If $\frac{\ln(\ln 2 |V_i|/|X_i|)}{\ln 2}$ is not an integer, we round m_i either to the ceiling or to the floor, depending on which one results in a smaller value of $f(m_i)$.

For now, we assume that we know $|W|$ and $|Y|$ in our computation of m_i. Later we will show how to estimate these values on the fly in the execution of each round of our protocol. Initially, $|X_1| (= |X|)$ is known. $|V_1|$ can be calculated from (2.16). Hence, the value of m_1 can be computed from (2.22). After that, we can estimate

$|Y_2|$, $|X_2|$, and $|V_2|$ based on (2.18), (2.19), and (2.16), respectively. From $|X_2|$ and $|V_2|$, we can calculate the value m_2. Following the same procedure, we can iteratively compute all values of m_i for $1 \leq i \leq K$.

We find it often happens that the m_i sequence has several consecutive zeros at the end, that is, $\exists p < K$, $m_i = 0$ for $i \in [p, K]$. In this case, we may be able to further optimize the value of m_p with a slight adjustment. We first explain the reason for $m_p = 0$: It costs some time for the reader to broadcast a filtering vector in phase one of the pth round. It is true that this filtering vector can reduce set Y_p, thereby reducing the frame size of phase two in the pth round. However, if the time cost of sending the filtering vector cannot be compensated by the time reduction of phase two, it will be better off to remove this filtering vector by setting $m_p = 0$. (This situation typically happens near the end of the m_i sequence because the number of unwanted tags in the remaining candidate set Y_p is already very small.) But if all values of m_i in the subsequent rounds (after m_p) are zeros, increasing m_p to a non-zero value m_p' may help reduce the transmission time of phase two of all subsequent rounds, and the total time reduction may compensate more than the time cost of sending those m_p' filtering vectors.

Consider the transmission time of these $(K - p + 1)$ rounds as a whole, denoted by $G(m_p', p)$. It is easy to derive

$$G(m_p', p) = \left(\frac{m_p'}{\ln 2} |X_p| + \frac{K - p + 1}{\ln 2} \left(\frac{|V_p|}{2^{m_p'}} + |W| \right) \right) t_s. \qquad (2.23)$$

To minimize $G(m_p', p)$, we have

$$m_p' = \begin{cases} 0 & \text{if } \gamma < 0 \\ \gamma & \text{if } \gamma \geq 0 \end{cases} \qquad (2.24)$$

where $\gamma = \frac{\ln(\ln 2(K-p+1)|V_p|/|X_p|)}{\ln 2}$. As a result, m_p is updated to m_p', while other m_i, $i \neq p$, remains unchanged.

Here, we give an example to illustrate how to calculate the values of m_i. Suppose $|X| = 5000$, $|Y| = 50{,}000$, $|W| = 500$, and $P_{REQ} = 0.001$, so $K = \lceil \frac{-\ln 0.001}{\ln 2} \rceil = 10$. Using (2.22), we can calculate the values from m_1 to m_{10}. The result is listed in Table 2.2. There is a sequence of zeros from m_7 to m_{10}. Thus, we can make an improvement using (2.24), and the optimized result is shown in Table 2.3.

Table 2.2 The initial values of m_i.

m_1	m_2	m_3	m_4	m_5	m_6	m_7	m_8	m_9	m_{10}
3	1	0	1	0	1	0	0	0	0

	m_1	m_2	m_3	m_4	m_5	m_6	m_7	m_8	m_9	m_{10}
Table 2.3 The optimized values of m_i.	3	1	0	1	0	1	2	0	0	0

2.3.7 Iterative Tag Search Protocol

Having calculated the values of m_i, we can present our iterative tag search protocol (ITSP) based on the generalized approach in Sect. 2.3.5. The protocol consists of K iterative rounds. Each round consists of two phases. Consider the ith round, where $1 \leq i \leq K$.

2.3.7.1 Phase One

The RFID reader constructs m_i filtering vectors for X_i using m_i hash functions. According to (2.12), we set the size L_{X_i} of each filtering vector as

$$L_{X_i} = \frac{1}{\ln 2} \times |X_i|. \tag{2.25}$$

The RFID reader then broadcasts those filtering vectors one by one. Once receiving a filtering vector, each tag in Y_i maps its ID to a bit in the filtering vector using the same hash function that the reader uses to construct the filter. The tag checks whether this bit is "1". If so, it remains a candidate tag; otherwise, it is excluded as a non-candidate tag and drops out of the search process immediately. The set of remaining candidate tags is Y_{i+1}.

If the filtering vectors are too long, the reader divides each vector into blocks of a certain length (e.g., 96 bits) and transmits one block after another. Knowing which bit it is mapped to, each tag only needs to record one block that contains its bit.

From (2.13), we know that the false-positive probability after using m_i filtering vectors is $(P_{FV})^{m_i} \approx (1/2)^{m_i}$. Therefore, $|Y_{i+1}| \approx |V_i| \times (P_{FV})^{m_i} + |W| \approx |V_i|/2^{m_i} + |W|$.

2.3.7.2 Phase Two

The reader broadcasts the frame size $L_{Y_{i+1}}$ of phase two to the tags, where

$$L_{Y_{i+1}} = \frac{1}{\ln 2} \left(|V_i|/2^{m_i} + |W| \right). \tag{2.26}$$

After receiving $L_{Y_{i+1}}$, each tag in Y_{i+1} randomly maps its ID to a slot in the time frame using a hash function and transmits a one-bit short response to the reader in that slot. Based on the observed state (busy or empty) of the slots in the time frame, the reader builds a filtering vector, which is used to filter non-candidates from X_i.

The overall transmission time of all K rounds in the ITSP is

$$T_{ITSP} = \sum_{i=i}^{K} (m_i \times L_{X_i} + L_{Y_{i+1}}) \times t_s. \qquad (2.27)$$

2.3.8   Cardinality Estimation

Recall from Sect. 2.3.6 that we must know the values of $|X_i|$, $|W|$, and $|V_i|$ to determine m_i, L_{X_i}, and $L_{Y_{i+1}}$. It is trivial to find the value of $|X_i|$ by counting the number of tags in the search result of the $(i-1)$th round. Meanwhile, we know $|V_i| \approx |V_{i-1}|/2^{m_{i-1}}$ and $|V_1| = |Y_1| - |W|$. Therefore, we only need to estimate $|W|$ and $|Y_1|$.

Besides serving as a filter, a filtering vector can also be used for cardinality estimation, a feature that is not exploited in [28]. Since no filtering vector is available at the very beginning, the first round of the ITSP should be treated separately: We may use the efficient cardinality estimation protocol ART [26] to estimate $|Y|$ (i.e., $|Y_1|$) if its value is not known at first. As for $|W|$, it is initially assumed to be $min\{|X|, |Y|\}$.

Next, we can take advantage of the filtering vector received by the reader in phase two of the ith ($i \geq 1$) round to estimate $|W|$ without any extra transmission expenditure. The estimation process is as follows: First, counting the actual number of "1" bits in the filtering vector, denoted as N_1^*, we know the actual false-positive probability of using this filtering vector, denoted by P_i^*, is

$$P_i^* = N_1^*/L_{Y_{i+1}}, \qquad (2.28)$$

because an arbitrary unwanted tag has a chance of N_1^* out of $L_{Y_{i+1}}$ to be mapped to a "1" bit, where $L_{Y_{i+1}}$ is the size of the vector. Meanwhile, we can record the number of tags in the search results before and after the ith round, i.e., $|X_i|$ and $|X_{i+1}|$, respectively. We have $|X_i| = |U_i| + |W|$, $|X_{i+1}| = |U_{i+1}| + |W|$, and $|U_{i+1}| \approx |U_i| \times P_i^*$. Therefore,

$$|W| \approx \frac{|X_{i+1}| - |X_i| \times P_i^*}{1 - P_i^*}. \qquad (2.29)$$

For the purpose of accuracy, we may estimate $|W|$ after every round, and obtain the average value.

2.3.9 Additional Filtering Vectors

Estimation may have error. Using the values of m_i and L_{Y_i} computed from estimated $|W|$ and $|Y_i|$, a direct consequence is that the actual false-positive ratio, denoted as P_T, can be greater than the requirement P_{REQ}. Fortunately, from (2.28), the reader is able to compute the actual false-positive ratio P_i^*, $1 \le i \le k$, of each filtering vector received in phase two of the ITSP. Thus, we have

$$P_T = \prod_1^K P_i^*. \tag{2.30}$$

If $P_T > P_{REQ}$, our protocol will automatically add additional filtering vectors to further filter X_{K+1} until $P_T \le P_{REQ}$ (as described in Sect. 2.3.4).

2.3.10 Hardware Requirement

ITSP cannot be supported by off-the-shelf tags that conform to the EPC Class-1 Gen-2 standard [9], whose limited hardware capability constrains the functions which can be supported. By our design, most of the ITSP protocol's complexity is on the reader side, but tags also need to provide certain hardware support. Besides the mandatory commands of C1G2 (e.g., Query, Select, and Read), in order for a tag to execute the ITSP protocol, we need a new command defined in the set of optional commands, asking each awake tag to listen to the reader's filtering vector, hash its ID to a certain slot of the vector for its bit value, keep silent and go sleep if the value is zero, and respond in a hashed slot (by making a transmission to make the channel busy) if the value is one. Note that the tag does not need to store the entire filtering vector, but instead only need to count to the slot it is hashed to, and retrieve the value (0/1) carried in that slot.

Hardware-efficient hash functions [1, 13, 22] can be found in the literature. A hash function may also be derived from the pseudo-random number generator required by the C1G2 standard. To keep the complexity of a tag's circuit low, we only use one uniform hash function $h(\cdot)$, and use it to simulate multiple independent hash functions: In phase one of the ith round, we use $h(\cdot)$ and m_i unique hash seeds $\{s_1, s_2, \cdots, s_{m_i}\}$ to achieve m_i independent hash outputs. Thus, a tag id is mapped to bit locations $(h(id \oplus s_1) \bmod L_{X_i})$, $(h(id \oplus s_2) \bmod L_{X_i})$, $\cdots$, $(h(id \oplus s_{m_i}) \bmod L_{X_i})$ in the m_i filtering vectors, respectively. Each hash seed, together with its corresponding filtering vector, will be broadcast to the tags. In phase two of the ith round, the reader generates a new hash seed s' and sends it to the tags. Each candidate tag in Y_{i+1} maps its id to the slot of index $\left(h(id \oplus s') \bmod L_{Y_{i+1}}\right)$, and then transmits a one-bit short response to the reader in that slot.

2.4   ITSP over Noisy Channel

So far the ITSP assumes that the wireless channel between the RFID reader and tags is reliable. Note that the CATS protocol does not consider channel error, either. However, it is common in practice that the wireless channel is far from perfect due to many different reasons, among which interference noise from nearby equipment, such as motors, conveyors, robots, wireless LAN's, and cordless phones, is a crucial one. Therefore, this section is to enhance ITSP by making it robust against noise interference.

2.4.1   ITSP with Noise on Forward Link

The reader transmits at a power level much higher than the tags (which after all backscatter the reader's signals in the case of passive tags). It has been shown that the reader may transmit more than one million times higher than tag backscatter [14]. Hence, the forward link (reader to tag) communication is more resilient against channel noise than the reverse link (tag to reader). To provide additional assurance against noise for forward link, we may use CRC code for error detection. The C1G2 standard requires the tags to support the computation of CRC-16 (16-bit CRC) [9], which therefore can also be adopted by future tags modified for ITSP. Each filtering vector built by the reader can be regarded as a combination of many small segments with fixed size of l_S bits (e.g., $l_S = 80$). For each segment, the reader computes its 16-bit CRC and appends it to end of that segment. Those segments are then concatenated and transmitted to tags. When a tag receives a filtering vector, it first finds the segment it hashes to and computes the CRC of that segment. If the calculated CRC matches the attached one, it will determine its candidacy by checking the bit in the segment which it maps to. For mismatching CRC, the tag knows that the segment has been corrupted, and it will remain as a candidate tag regardless of the value of the bit which it maps to.

Suppose we let $l_S = 80$, then

$$L_{X_i} = \frac{\frac{1}{\ln 2} \times |X_i|}{l_S} \times (l_S + 16) = \frac{1.2|X|}{\ln 2}. \tag{2.31}$$

We assume the probability that the noise corrupts each segment is P_S (P_S is expected to be very small as explained above). A corrupted segment can be thought as consisting of all "1"s. Hence, the false-positive probability for a filtering vector sent by reader, denoted by P_{RT}, is roughly

$$P_{RT} \approx \frac{\frac{Lx_i}{96} \times P_S \times l_S + \frac{Lx_i}{96} \times (1 - P_S) \times l_S \times P_{FV}}{\frac{Lx_i}{96} \times l_S} \tag{2.32}$$

$$= \frac{1 + P_S}{2}.$$

We can also get

$$|Y_{i+1}| \approx |V_i| \times (P_{RT})^{m_i} + |W| \tag{2.33}$$

and now (2.20) can be rewritten as

$$f(m_i) = \frac{t_s}{\ln 2} \left(1.2 m_i |X_i| + \left(\frac{1 + P_{RT}}{2} \right)^{m_i} |V_i| + |W| \right). \tag{2.34}$$

Therefore, $f(m_i)$ is optimized when

$$m_i = \frac{\ln[(\ln 2 - \ln(1 + P_{RT}))|V_i|/1.2|X_i|]}{\ln 2 - \ln(1 + P_{RT})}. \tag{2.35}$$

2.4.2 ITSP with Noise on Reverse Link

Now let us study the noise on the reverse link and its effect on the ITSP. Since the backscatter from a tag is much weaker than the signal transmitted by the reader, the reverse link is more likely to be impacted by noise.

First, channel noise may corrupt a would-be empty slot into a busy slot. The original empty slot is supposed to be translated into a "0" bit in the filtering vector by the reader; if a candidate tag is mapped to that bit, it is ruled out immediately. However, if that slot is corrupted and becomes a busy slot, the corresponding bit turns into "1"; a tag mapped to that bit will remain a candidate tag, thereby increasing the false-positive probability of the filtering vector.

Second, noise may also occur during a busy slot. Although the noise and the transmissions from tags may partially cancel each other in a slot if they happen to reach the reader in opposite phase, it is extremely unlikely that they will exactly eliminate each other. As long as the reader can still detect some energy, regardless of its source (it may even come from the noise), that slot will be correctly determined as a busy slot, and the corresponding bit in the filtering vector is set to "1" just as it is supposed to be. However, if we take the propagation path loss, including reflection loss, attenuation loss, and spreading loss [11], into account, there is still a chance that a busy slot may not be detected by the reader. This may happen in a time-varying channel where the reader may fail in receiving a tag's signal during a deeply faded slot when the tag transmits. We stress that this is not a problem unique to ITSP, but all protocols that require communications from tags to readers will suffer from this problem if it happens that the reader cannot hear the tags. ITSP is not robust against this type of error. But there exists ways to alleviate this problem—for instance, each filtering vector from tags to the reader is transmitted twice. As long as a slot is busy in one of two transmissions, the slot is considered to be busy.

Next, we will investigate the reverse link with noise interference for ITSP under two error models.

2.4.2.1 ITSP Under Random Error Model (ITSP-rem)

The random error model is characterized by a parameter called error rate P_{ERR}, which means every slot independently has a probability P_{ERR} to be corrupted by the noise. Influencing by the channel noise, the reader can detect more busy slots as some empty slots turn into busy ones, which raises the false-positive probability of phase-two filtering vectors. Suppose the frame size of phase two in a certain round is l, the original number of busy slots is about $l \times P_{FV} \approx l/2$. At the reader's side, however, the number of busy slots averagely increases to $l/2 + l/2 \times P_{ERR} = \frac{(1+P_{ERR}) \times l}{2}$. After encoding the slot status into a filtering vector, the false-positive probability of that filtering vector is

$$P'_{FV} \approx \frac{\frac{(1+P_{ERR}) \times l}{2}}{l} = \frac{1 + P_{ERR}}{2}. \tag{2.36}$$

To satisfy the false-positive ratio requirement, $\left(P'_{FV}\right)^K \leq P_{REQ}$ should hold. Therefore, the search process of ITSP-rem contains at least

$$K = \left\lceil \frac{\ln P_{REQ}}{\ln[(1 + P_{ERR})/2]} \right\rceil \tag{2.37}$$

rounds. Also, we can derive

$$\begin{aligned} |X_{i+1}| &\approx |U_i| \times P'_{FV} + |W| \\ &\approx |U_i|(1 + P_{ERR})/2 + |W|. \end{aligned} \tag{2.38}$$

With K, $|X_i|$, $|Y_i|$ and m_i, $1 \leq i \leq K$, the search time of ITSP-rem can be calculated using (2.31) (2.26) (2.27).

2.4.2.2 ITSP Under Burst Error Model (ITSP-bem)

In telecommunication, a burst error is defined as a consecutive sequence of received symbols, where the first and last symbols are in error, and there exists no continuous subsequence of m (m is a specified parameter called the guard band of the error burst) correctly received symbols within the error burst [10]. A burst error model describes the number of bursts during an interval and the number of incorrect symbols in each burst error, which differs greatly from the random error model.

According to the burst error model presented in [6], both the number of bursts in an interval and the number of errors in each burst have Poisson distributions. Assume the expected number of bursts in an l-bit interval is η, the probability distribution function for the number of bursts can be expressed as

$$h(x) = \sum_{i=0}^{\infty} \frac{\eta^i}{i!} e^{-\eta} \delta_{xi}, \tag{2.39}$$

where δ_{xi} is the Kronecker delta function [18]. Meanwhile, if the mean value of errors due to a burst in the l bits is τ, then the probability distribution function of the number of error is given by

$$g(y) = \sum_{j=0}^{\infty} \frac{\tau^j}{j!} e^{-\tau} \delta_{yj}. \tag{2.40}$$

Therefore, the probability of having w errors in an interval of l bits is

$$P_l(w) = e^{-\eta} \frac{\tau^w}{w!} \sum_{i=0}^{\infty} \frac{i^w}{i!} \eta^i e^{-i\tau}. \tag{2.41}$$

In other words, for a frame with l slots, the probability that w slots will be corrupted by the burst noise is $P_l(w)$.

Now we evaluate the ITSP under the burst error model, denoted as ITSP-bem. Given a filtering vector with size of l-bit, recall from (2.41) that the probability of having w errors in this l-bit vector is $P_l(w)$. In this case, each original "0" bit has a probability $\frac{w}{l}$ to be corrupted by the errors, and becomes a "1" bit. Consequently, the false-positive probability of the filtering vector is expected to be:

$$P'_{FV} \approx \frac{1}{2} + \frac{1}{2} \sum_{w=0}^{l} P_l(w) \times \frac{w}{l}. \tag{2.42}$$

After obtaining the value of P'_{FV}, the ITSP-bem can use (2.37), (2.38), to determine the values of other necessary parameters.

2.5 Performance Evaluation

2.5.1 Performance Metric

We compare our protocol ITSP with CATS [28], the polling protocol (Sect. 2.2.2), the optimal DFSA (dynamic frame slotted ALOHA), and a tag identification protocol with collision recovery [15], denoted as CR, which identifies 4.8 tags per slot on average, about 13 times the speed of the optimal DFSA. For ITSP and CATS, their Bloom filters (or filtering vectors) constitute most of the overall transmission overhead, while other transmission cost, such as transmission of hash seeds, is comparatively negligible. Both protocols need to estimate the number of tags in the system, $|Y|$, as a pre-protocol step. According to the results presented in [28], the time for estimating $|Y|$ takes up less than 2% of the total execution time of CATS. Hence, we do not count the estimation time of $|Y|$ in the simulation results because it is relatively small and does not affect fair comparison as both protocols need it. Consequently, the key metric concerning the time efficiency is the total size of Bloom filters or filtering vectors, and then (2.8) can be used for calculating the search time required by CATS, while (2.27) for ITSP.

After the search process is completed, we will calculate the false -positive ratio P_{FP} using $P_{FP} = \frac{|W^*-W|}{|X-W|}$, where W^* is the set of tags in the search result and W is the actual set of wanted tags in the coverage area. P_{FP} will be compared with P_{REQ} to see whether the search result meets the false -positive ratio requirement.

2.5.2 Performance Comparison

We evaluate the performance of our protocol and compare it with the CATS protocol. In the first set of simulations, we set $P_{REQ} = 0.001$, fix $|Y| = 50,000$, vary $|X|$ from 5000 to 640,000, and let $R_{INTS} = 0.1, 0.3, 0.5, 0.7, 0.9$. In the second set of simulations, we set $P_{REQ} = 0.001$, fix $|X| = 10,000$, vary $|Y|$ from 1250 to 40,000 to investigate the scalability of ITSP with tag population from a large range, and let $R_{INTS} = 0.1, 0.3, 0.5, 0.7, 0.9$. For simplicity, we assume $t_{id} = 96t_s$, and $t_l = 137t_s$, in which a 9-bit QueryAdjust or a 4-bit QueryRep command, a 96-bit ID and two 16-bit random numbers can be transmitted. Tables 2.4 and 2.5 show the number of t_s slots needed by the protocols under different parameter settings. Each data point in these tables or other figures/tables in the rest of the section is the average of 500 independent simulation runs with $\pm 5\%$ or less error at 95% confidence level.

From the tables, we observe that when R_{INTS} is small (which means $|W|$ is small), the ITSP performs much better than the CATS protocol. For example, in Table 2.4, when $R_{INTS} = 0.1$, the ITSP reduces the search time of the CATS protocol by as much as 90.0%. As we increase R_{INTS} (which implies larger $|W|$), the gap between the performance of the ITSP and the performance of the CATS gradually shrinks.

Table 2.4 Performance comparison of tag search protocols. CR means a tag identification protocol with collision recovery techniques. $|Y| = 50{,}000$, $P_{REQ} = 0.001$

$|X|$	ITSP (R_{INTS})					CATS	Polling	CR
	0.1	0.3	0.5	0.7	0.9			
5,000	61,463	96,989	105,828	108,346	124,553	126,370	485,000	1,427,083
10,000	108,017	145,553	206,709	199,586	231,236	238,313	970,000	1,427,083
20,000	185,204	255,898	335,426	397,462	403,954	447,772	1,940,000	1,427,083
40,000	304,767	467,433	512,156	598,718	678,066	837,837	3,880,000	1,427,083
80,000	414,686	590,150	656,426	721,347	721,347	1,560,259	7,760,000	1,427,083
160,000	472,677	630,669	721,347	721,347	721,347	2,889,689	15,520,000	1,427,083
320,000	529,835	668,794	721,347	721,347	721,347	5,317,715	31,040,000	1,427,083
640,000	573,270	696,015	721,347	721,347	721,347	10,533,732	62,080,000	1,427,083

Table 2.5 Performance comparison of tag search protocols. CR means a tag identification protocol with collision recovery techniques. $|X| = 10{,}000$, $P_{REQ} = 0.001$

$|Y|$	ITSP (R_{INTS})					CATS	Polling	CR
	0.1	0.3	0.5	0.7	0.9			
1,250	13,047	17,364	18,033	18,033	18,033	164,589	970,000	35,677
2,500	24,289	33,337	36,067	36,067	36,067	175,960	970,000	71,354
5,000	42,835	62,862	68,528	72,134	72,134	190,387	970,000	142,708
10,000	73,909	109,281	119,022	137,056	144,269	204,814	970,000	285,417
20,000	95,833	132,546	169,065	167,713	192,960	219,241	970,000	570,833
40,000	111,904	152,606	174,926	228,215	232,904	233,668	970,000	1,141,667

In particular, the CATS performs poorly when $|X| \geq |Y|$. But the ITSP can work efficiently in all cases. In addition, the ITSP is also much more efficient than the polling protocol, and any tag identification protocol with/without CR techniques. Even in the worst case, the ITSP only takes about half of the execution time of a tag identification protocol with CR techniques. (Note that the identification process actually takes much more time since the throughput 4.8 tags per slot may not be achievable in practical and the duration of each slot is longer.) In practice, the wanted tags may be spatially distributed in many different RFID systems (e.g., warehouses in the example we use in the introduction), and thus R_{INTS} can be small. The ITSP is a much better protocol for solving the tag search problem in these practical scenarios.

Another performance issue we want to investigate is the relationship between the search time and P_{REQ}. The polling protocol, DFSA, and CR do not have false positive. Our focus will be on ITSP and CATS. We set $|X| = 5000, 20{,}000$ or $80{,}000$, $|Y| = 50{,}000$, vary R_{INTS} from 0.1 to 0.9, and vary P_{REQ} from 10^{-6} to 10^{-2}. Figure 2.4 compares the search times required by the CATS and the ITSP under different false -positive ratio requirements. Generally speaking, the gap between the search time required by the ITSP and the search time by the CATS keeps getting larger with the decrease of P_{REQ}, particularly when R_{INTS} is small. For example, in

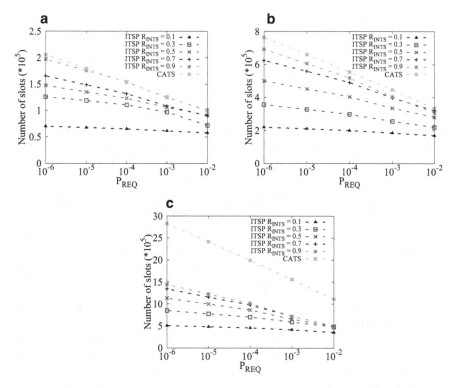

Fig. 2.4 Relationship between search time and P_{REQ}. Parameter setting: $|Y| = 50,000$; (**a**) $|X| = 5000$, (**b**) $|X| = 20,000$, (**c**) $|X| = 80,000$

Fig. 2.4c, when $P_{REQ} = 10^{-2}$ and $R_{INTS} = 0.1$, the search time by the ITSP is about one third of the time by the CATS; when we reduce P_{REQ} to 10^{-6}, the time by the ITSP becomes about one fifth of the time by the CATS. The reason is as follows: When R_{INTS} is small, $|W|$ is small and most tags in X and Y are non-candidates. After several ITSP rounds, as many non-candidates are filtered out iteratively, the size of filtering vectors decreases exponentially and therefore subsequent ITSP rounds do not cause much extra time cost. This merit makes the ITSP particularly applicable in cases where the false -positive ratio requirement is very strict, requiring many ITSP rounds. On the contrary, the CATS protocol does not have this capability of exploiting low R_{INTS} values.

2.5.3   False -Positive Ratio

Next, we examine whether the search results after execution of the ITSP will indeed meet the requirement of P_{REQ}. In this simulation, we set the false-positive ratio requirement based on the following formula:

$$P_{REQ} \leq \frac{|W|}{\lambda\,(|X| - |W|)}, \tag{2.43}$$

where λ is a constant. We use an example to give the rationale: Consider an RFID system with $|X| = 20{,}000$. If $|W| = 10{,}000$, $P_{REQ} = 0.01$ may be good enough because the number of false positives is about $(|X| - |W|) \times P_{REQ} = 100$, which is much fewer than $|W|$. However, if $|W| = 10$, $P_{REQ} = 0.01$ may become unacceptable since $(|X| - |W|) \times P_{REQ} \approx 200 \gg |W|$. Therefore, it is desirable to set the value of P_{REQ} such that the number of false positives in the search result is much smaller than $|W|$, namely $(|X| - |W|) \times P_{REQ} \leq \frac{1}{\lambda}|W|$. Let $\lambda = 10$ and we test the ITSP under three different parameter settings:

(1) $|X| = 5000$, $|Y| = 50{,}000$, and R_{INTS} varies from 0.1 to 0.9, i.e., $|W|$ varies from 500 to 4500. $P_{REQ} \leq \frac{500}{10 \times (5000 - 500)} \approx 0.01111$. We set $P_{REQ} = 10^{-2}$.
(2) $|X| = 20{,}000$, $|Y| = 50{,}000$, and R_{INTS} varies from 0.01 to 0.9, i.e., $|W|$ varies from 200 to 18,000. $P_{REQ} \leq \frac{200}{10 \times (20{,}000 - 200)} \approx 0.00101$. We set $P_{REQ} = 10^{-3}$.
(3) $|X| = 80{,}000$, $|Y| = 50{,}000$, and R_{INTS} varies from 0.01 to 0.9, i.e., $|W|$ varies from 500 to 45,000. $P_{REQ} \leq \frac{500}{10 \times (80{,}000 - 500)} \approx 0.00063$. We set $P_{REQ} = 10^{-4}$.

For each parameter setting, we repeat the simulation 500 times to obtain the average false -positive ratio.

Figure 2.5 shows the simulation results. In (a), (b), and (c), we can see that the average P_{FP} is always smaller than the corresponding P_{REQ}. Hence, the search results using the ITSP meet the prescribed requirement of false -positive ratio in the average sense.

If we look into the details of individual simulations, we find that a small fraction of simulation runs have P_{FP} beyond P_{REQ}. For example, Fig. 2.6 depicts the results of 500 runs with $|X| = 5000$, $|Y| = 50{,}000$, $|W| = 500$, and $P_{REQ} = 10^{-2}$. There are about 5 % runs having $P_{FP} > P_{REQ}$, but that does not come as a surprise because the false -positive ratio in the context of filtering vectors (ITSP) or Bloom filters (CATS) is defined in a probability way: The probability for each tag in $X - W$ to be misclassified as one in W is no greater than P_{REQ}. This probabilistic definition enforces a requirement P_{REQ} in an average sense, but not absolutely for each individual run.

2.5.4   Performance Evaluation Under Channel Error

2.5.4.1   Performance of ITSP-rem and ITSP-bem

We evaluate the performance of ITSP-rem and ITSP-bem. To simulate the error rate P_{ERR} in ITSP-rem, we employ a pseudo-random number generator, which generates random real numbers uniformly in the range [0, 1]. If a bit in the filtering vector is "0" and the generated random number is in [0, P_{ERR}], that bit is flipped to "1". P_S can be simulated in a similar way. As for the burst error in ITSP-bem, we first

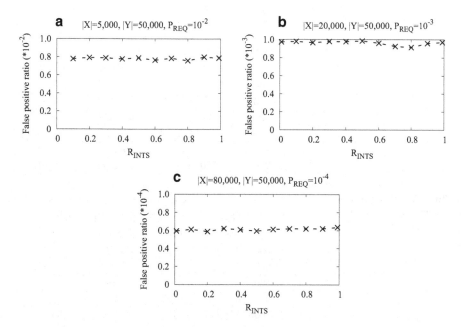

Fig. 2.5 False -positive ratio after running the ITSP

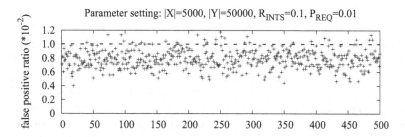

Fig. 2.6 False -positive ratio by the ITSP of 500 runs

calculate the values of $P_l(w)$ with different w for a given l. Then each w is assigned with a non-overlapping range in [0, 1], whose length is equal to the value of $P_l(w)$. For each interval, we generate a random number and check which range the number locates, thereby determining the number of errors in that interval.

We set $P_{REQ} = 0.001$, $P_S = 0.01$, and $R_{INTS} = 0.1, 0.5, 0.9$, respectively. The values of $|X|$ and $|Y|$ are the same as those in Tables 2.4 and 2.5. l_s is set to 80 bits and a 16-bit CRC is appended to each segment on forward link for integrity check. For ITSP-rem, we consider two cases with $P_{ERR} = 5\%$ and 10%, respectively. For ITSP-bem, the prescribed parameters are set to be: $\eta = 0.135$, $\tau = 7.10$ with each interval to be 96 bits [6].

Table 2.6 Performance comparison. $|Y| = 50,000$, $R_{INTS} = 0.1$, $P_{REQ} = 0.001$

$|X|$	ITSP	ITSP-rem $P_{ERR} = 5\%$	$P_{ERR} = 10\%$	ITSP-bem
5,000	61,463	74,288	75,812	72,144
10,000	108,017	129,995	133,022	125,779
20,000	185,204	241,026	247,824	238,962
40,000	304,767	361,242	398,198	358,361
80,000	414,686	441,365	458,433	437,256
160,000	472,677	504,565	545,338	499,058
320,000	529,835	567,403	630,174	560,456
640,000	573,270	626,379	690,400	618,913

Table 2.7 Performance comparison. $|Y|=50,000$, $R_{INTS}=0.5$, $P_{REQ} = 0.001$

$|X|$	ITSP	ITSP-rem $P_{ERR} = 5\%$	$P_{ERR} = 10\%$	ITSP-bem
5,000	105,828	160,481	166,469	153,838
10,000	206,709	211,513	221,771	210,805
20,000	335,426	371,974	391,983	370,557
40,000	512,156	577,305	617,196	577,305
80,000	656,426	735,592	789,874	735,592
160,000	721,347	793,482	865,617	793,482
320,000	721,347	793,482	865,617	793,482
640,000	721,347	793,482	865,617	793,482

Tables 2.6, 2.7, 2.8, 2.9, 2.10, and 2.11 show the number of t_s slots needed under each parameter setting. The second column presents the results of ITSP when the channel is perfectly reliable. The third and fourth columns present the results of ITSP-rem with an error rate of 5 % or 10 %. The fifth column presents the results of ITSP-bem. It is not surprising that the search process under noisy channel generally takes more time due to the use of CRC and the higher false-positive probability of filtering vectors, and the execution time of the ITSP-rem is usually longer in a channel with a higher error rate. An important positive observation is that the performance of ITSP gracefully degrades in all simulations. The increase in execution time for both ITSP-rem and ITSP-bem is modest, compared to ITSP with a perfect channel. For example, even when the error rate is 10 %, the execution time of ITSP-rem is about 10–30 % higher than that of ITSP. This modest increase demonstrates the practicality of our protocol under noisy channel.

2.5.4.2 False-Positive Ratio of ITSP-rem and ITSP-bem

We use the same parameter settings in Sect. 2.5.3 to examine the accuracy of search results by ITSP-rem and ITSP-bem. Meanwhile, for ITSP-rem, we set $P_{ERR} = 5\%$ or 10 %. For ITSP-bem, the required input parameter setting is $\eta = 0.135$ and $\tau = 7.10$, with each 96-bit interval. Simulation results are delineated in Fig. 2.7,

Table 2.8 Performance
comparison. $|Y|$=50,000,
R_{INTS}=0.9, $P_{REQ} = 0.001$

$|X|$	ITSP	ITSP-rem		ITSP-bem
		$P_{ERR} = 5\%$	$P_{ERR} = 10\%$	
5,000	124,553	156,041	163,718	155,972
10,000	231,236	275,394	290,493	275,256
20,000	403,954	454,929	486,150	454,929
40,000	678,066	752,753	814,890	752,753
80,000	721,347	793,482	865,617	793,482
160,000	721,347	793,482	865,617	793,482
320,000	721,347	793,482	865,617	793,482
640,000	721,347	793,482	865,617	793,482

Table 2.9 Performance
comparison. $|X|$=10,000,
R_{INTS}=0.1, $P_{REQ} = 0.001$

$|Y|$	ITSP	ITSP-rem		ITSP-bem
		$P_{ERR} = 5\%$	$P_{ERR} = 10\%$	
1,250	13,047	14,868	15,898	14,174
2,500	24,289	26,626	28,617	25,283
5,000	42,835	46,994	50,863	44,393
10,000	73,909	76,807	84,135	75,983
20,000	95,833	103,255	106,693	102,121
40,000	111,904	133,043	137,348	130,382

Table 2.10 Performance
comparison. $|X|$=10,000,
R_{INTS}=0.5, $P_{REQ} = 0.001$

$|Y|$	ITSP	ITSP-rem		ITSP-bem
		$P_{ERR} = 5\%$	$P_{ERR} = 10\%$	
1,250	18,033	19,837	21,640	19,837
2,500	36,067	39,674	43,280	39,674
5,000	68,528	77,021	82,448	77,021
10,000	119,022	134,208	143,261	134,208
20,000	169,065	202,891	212,105	202,467
40,000	174,926	214,563	224,227	213,970

Table 2.11 Performance
comparison. $|X|$=10,000,
R_{INTS}=0.9, $P_{REQ} = 0.001$

$|Y|$	ITSP	ITSP-rem		ITSP-bem
		$P_{ERR} = 5\%$	$P_{ERR} = 10\%$	
1,250	18,033	19,837	21,640	19,837
2,500	36,067	39,674	43,280	39,674
5,000	72,134	79,348	86,561	79,348
10,000	144,269	158,696	173,123	158,696
20,000	192,960	217,245	232,272	217,245
40,000	232,904	261,277	277,300	261,173

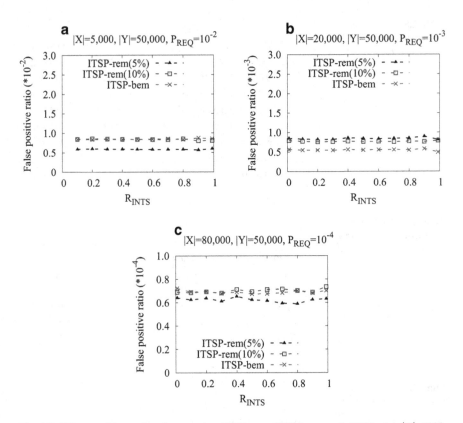

Fig. 2.7 False -positive ratio after running ITSP-rem, ITSP-bem, and CATS. (**a**) $|X|$=5000, $|Y|$=50,000, $P_{REQ} = 10^{-2}$, (**b**) $|X|$=20,000, $|Y|$=50,000, $P_{REQ} = 10^{-3}$, (**c**) $|X|$=80,000, $|Y|$=50,000, $P_{REQ} = 10^{-4}$

where the error rate is given between the parentheses after ITSP-bem. Clearly, the false-positive ratio in the search results after executing ITSP-rem or ITSP-bem is always within the bound of P_{REQ}. These results confirm that the false-positive ratio requirement is met under noisy channel.

2.5.4.3   Signal Loss Due to Fading Channel

We consider the scenario of a time-varying channel in which it may happen that a signal from a tag is not received by the reader in a deep fading slot. Although we consider this condition is relatively rare in an RFID system that is configured to work stably, we acknowledge in Sect. 2.4.2 that ITSP (or CATS) is not robust against this type of error. However, the problem can be alleviated by the tags transmitting each filtering vector twice. Figure 2.8 shows the simulation results under parameters $|X| = 10000$, $|Y| = 5000$, $|W| = 500$, and $P_{REQ} = 0.01$. The horizontal axis

Fig. 2.8 False negatives due to signal loss in time-varying channel

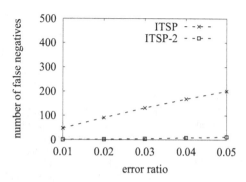

shows the error rate, which is defined as the fraction of slots in deep fading, causing complete signal loss. ITSP-2 denotes the approach of transmitting each filtering vector from tags to the reader twice. When a wanted tag in W is not identified, we call it a false negative. The simulation results show that ITSP incurs significant false negatives when the error rate becomes large. For example, when the error rate is 2 %, the average number of false negatives is 90.7. ITSP-2 works very well in reducing this number. When the error rate is 2 %, its number of false negatives is just 1.95.

2.6 Summary

This chapter discusses the tag search problem in large-scale RFID systems. We present an iterative tag search protocol (ITSP) that improves time efficiency and eliminates the limitation of prior solutions. Moreover, we extend the ITSP to work under noisy channel. The main contributions of our work are summarized as follows: (1) The iterative method of ITSP based on filtering vectors is very effective in reducing the amount of information to be exchanged between tags and the reader, and consequently saves time in the search process; (2) the ITSP performs much better than the existing solutions; (3) the ITSP works well under all system conditions, particularly in situations of $|X| \gg |Y|$ when CATS works poorly; (4) the ITSP is improved to work effectively under noisy channel.

References

1. Bogdanov, A., Leander, G., Paar, C., Poschmann, A., Robshaw, M.J.B., Seurin, Y.: Hash functions and RFID tags: mind the gap. In: Proceedings of CHES, pp. 283–299 (2008)
2. Broder, A., Mitzenmacher, M.: Network applications of bloom filters: a survey. Internet Math. 1(4), 485–509 (2003)
3. Castiglione, P., Ricciato, F., Popovski, P.: Pseudo-random Aloha for inter-frame soft combining in RFID systems. In: Proceedings of IEEE DSP, pp. 1–6 (2013)

4. Cha, J.R., Kim, J.H.: Dynamic framed slotted ALOHA algorithms using fast tag estimation method for RFID systems. In: Proceedings of IEEE Consumer Communications and Networking Conference (CCNC) (2006)
5. Choi, J., Lee, C.: A cross-layer optimization for a LP-based multi-reader coordination in RFID systems. In: Proceedings of IEEE GLOBECOM, pp. 1–5 (2010)
6. Cornaglia, B., Spini, M.: New statistical model for burst error distribution. Eur. Trans. Telecommun. **7**, 267–272 (1996)
7. Dan, L., Wei, P., Wang, J., Tan, J.: TFDMA: a scheme to the RFID reader collision problem based on graph coloration. In: Proceedings of IEEE SOLI, pp. 502–507 (2008)
8. Eom, J., Lee, T.: Accurate tag estimation for dynamic framed-slotted ALOHA in RFID systems. In: Proceedings of IEEE Communication Letters, pp. 60–62 (2010)
9. EPC Radio-Frequency Identity Protocols Class-1 Gen-2 UHF RFID Protocol for Communications at 860MHz-960MHz, EPCglobal. Available at http://www.epcglobalinc.org/uhfclg2 (2011)
10. Federal Standard 1037C. Available at http://www.its.bldrdoc.gov/fs-1037/fs-1037c.htm (1996)
11. Fletcher, R., Marti, U.P., Redemske, R.: Study of UHF RFID signal propagation through complex media. In: IEEE Antennas and Propagation Society International Symposium, vol. 1B, pp. 747–750 (2005)
12. Fyhn, K., Jacobsen, R.M., Popovski, P., Scaglione, A., Larsen, T.: Multipacket reception of passive UHF RFID tags: a communication theoretic approach. IEEE Trans. Signal Process. **59**(9), 4225–4237 (2011)
13. Guo, J., Peyrin, T., Poschmann, A.: The PHOTON family of lightweight Hash functions. In: Proceedings of CRYPTO, pp. 222–239 (2011)
14. NIST: RFID Communication and Interference. White paper, Grand Prix Application Series (2007)
15. Kaitovic, J., Rupp, M.: Improved physical layer collision recovery receivers for RFID readers. In: Proceedings of IEEE RFID, pp. 103–109 (2014)
16. Kaitovic, J., Langwieser, R., Rupp, M.: A smart collision recovery receiver for RFIDs. EURASIP J. Embed. Syst. **2013**, 1–19 (2013)
17. Kang, Y., Kim, M., Lee, H.: A hierarchical structure based reader anti-collision protocol for dense RFID reader networks. In: Proceedings of ICACT, pp. 164–167 (2011)
18. Kronecker delta. Available at http://en.wikipedia.org/wiki/Kronecker_delta
19. Lee, S., Joo, S., Lee, C.: An enhanced dynamic framed slotted ALOHA algorithm for RFID tag identification. In: Proceedings of IEEE MobiQuitous (2005)
20. Nguyen, C.T., Hayashi, K., Kaneko, M., Popovski, P., Sakai, H.: Probabilistic dynamic framed slotted ALOHA for RFID tag identification. Wirel. Pers. Commun. **71**, 2947–2963 (2013)
21. Onat, I., Miri, A.: A tag count estimation algorithm for dynamic framed ALOHA based RFID MAC protocols. In: Proceedings of IEEE ICC, pp. 1–5 (2011)
22. O'Neill, M.: Low-cost SHA-1 hash function architecture for RFID tags. In: Proceedings of RFIDSec (2008)
23. Qiao, Y., Li, T., Chen, S.: One memory access bloom filters and their generalization. In: Proceedings of IEEE INFOCOM, pp. 1745–1753 (2011)
24. Ricciato, F., Castiglione, P.: Pseudo-random ALOHA for enhanced collision-recovery in RFID. IEEE Commun. Lett. **17**(3), 608–611 (2013)
25. Schoute, F.C.: Dynamic frame length ALOHA. IEEE Trans. Commun. **31**, 565–568 (1983)
26. Shahzad, M., Liu, A.: Every bit counts - fast and scalable RFID estimation. In: Proceedings of ACM Mobicom (2012)
27. Stefanovic, C., Popovski, P.: ALOHA random access that operates as a rateless code. IEEE Trans. Commun. **61**(11), 4653–4662 (2013)
28. Zheng, Y., Li, M.: Fast tag searching protocol for large-scale RFID systems. IEEE/ACM Trans. Networking **21**(3), 924–934 (2012)

Chapter 3
Lightweight Anonymous RFID Authentication

This chapter describes on lightweight RFID anonymous authentication. The widespread use of RFID tags raises a privacy concern: They make their carriers trackable. To protect the privacy of the tag carriers, we need to invent new mechanisms that keep the usefulness of tags while doing so anonymously. Many tag applications such as toll payment require authentication. Since low-cost tags have extremely limited hardware resource, an asymmetric design principle is adopted to push most complexity to more powerful RFID readers. Instead of implementing complicated and hardware-intensive cryptographic hash functions, our authentication protocol only requires tags to perform several simple and hardware-efficient operations to generate dynamic tokens for anonymous authentication. The theoretic analysis and randomness tests demonstrate that our protocol can ensure the privacy of the tags. Moreover, our protocol reduces the communication overhead and online computation overhead to $O(1)$ per authentication for both tags and readers, which compares favorably with the prior art.

The rest of this chapter is organized as follows. Section 3.1 describes the system model and the security model. Section 3.2 gives the related work. Sections 3.3–3.5 present three anonymous RFID authentication protocols. Security analysis is performed in Sect. 3.6. Section 3.7 provides the numerical evaluation. Section 3.8 gives the summary.

3.1 System Model and Security Model

3.1.1 System Model

Consider a hierarchical distributed RFID system as shown in Fig. 3.1. Each tag is pre-installed with some keys for authentication. The readers are deployed at

© Springer International Publishing AG 2016
M. Chen, S. Chen, *RFID Technologies for Internet of Things*,
Wireless Networks, DOI 10.1007/978-3-319-47355-0_3

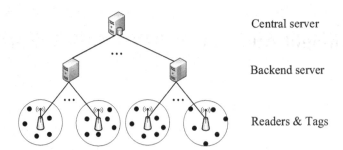

Central server

Backend server

Readers & Tags

Fig. 3.1 A hierarchical distributed RFID system

chosen locations, responsible for authenticating tags entering their coverage areas. In addition, the readers at each location are connected to a backend server, serving as a supplement to provide more storage and computation resources. All backend servers are further connected to the central server, where every tag' keys are stored. Any authorized backend server can fetch the tags' keys from the central server. Since the keys of each tag are only stored at the central server, they are synchronized from the view of different backend servers. Moreover, the high speed links connecting the central server, backend servers, and readers make the latency of transmitting small authentication data negligible. Therefore, a reader, its connected backend server, and the central server can be thought as single entity, and will be used interchangeably.

In this chapter, we focus on low-cost RFID tags, particularly passive backscatter tags that are ubiquitously used nowadays. The simplicity of these tags contributes to their low prices, which in turn restricts their computation, communication, and storage capabilities. In contrast, the readers, which are not needed in a large quantity as tags do, can have much richer resource. Moreover, the backend server can provide the readers with extra resource when necessary. The communication between a reader and a tag works in the request-and-response mode. The reader initiates the communication by sending a request. Upon receiving the request, the tag makes an appropriate transmission in response. We divide the transmissions between the readers and tags into two types: (1) Invariant transmissions contain the content that is invariant between any tag and any reader, such as the beacon transmission from a reader which informs the incoming tag of what to do next. (2) Variant transmissions contain the content that may vary for different tags or the same tag at different times, such as the exchanged data for anonymous authentication.

3.1.2 Security Model

Threat Model An adversary may eavesdrop on any wireless transmissions made between the tags and the readers. In addition, the adversary may plant unauthorized readers at chosen locations, which communicate with passing tags and try to identify the tag carriers. However, such unauthorized readers have no access to the backend servers or the central server since the servers will authenticate the readers before granting access permissions. In the sequel, a reader without further notation means an authorized one by default. Moreover, we assume that the adversary may compromise some tags and obtain their keys, but it cannot compromise any authorized readers.

Anonymous Model The anonymous model requires that all variant transmissions must be indistinguishable by the adversary, meaning that (1) any variant transmission in the protocol should not carry a fixed value that is unchanged across multiple authentications, and (2) the transmission content should appear totally random and unrelated across different authentications to any eavesdropper that captures the transmissions. Therefore, no adversary will have a non-negligible advantage in successfully guessing the next variant transmission of a tag based on the previous transmissions [9].

Notations used in the chapter are given in Table 3.1 for quick reference.

Table 3.1 Notations

Symbols	Descriptions
n	Number of tags in the distributed RFID system
n_t	Total number of tokens used by the tags
R	An RFID reader
t	An RFID tag
idx	A tag index $(1 \leq i \leq n)$
tk	A token for authentication
ic	An indicator
$tk[i]$	The ith bit in the token tk
$ic[i]$	The ith bit in the indicator ic
t_i	The ith $(1 \leq i \leq n)$ tag in the system
KT	Key table stored in the database
$KT[i]$	The keys of tag t_i stored in the key table
tk_i^j	The jth token stored by tag t_i
pt_i	The token index stored by tag t_i
u	Number of base tokens stored by each tag
v	Number of base indicators stored by each tag
bt_i^j	The jth $(1 \leq j \leq u)$ base token of tag t_i
bi_i^j	The jth $(1 \leq j \leq v)$ base indicator of tag t_i
ic_i	The current indicator of tag t_i
a	Length of each token/base token
b	Length of each indicator/base indicator
ρ	Memory utilization ratio of the hash table

3.2 Related Work

Prior work on anonymous authentication can be generally classified into two
categories: non-tree-based and tree-based.

3.2.1 Non-tree-Based Protocols

The hash-lock [20] leverages random hash values for anonymous authentication.
After receiving an authentication request from a reader, a tag sends back $(r, id \oplus f_k(r))$, where r is a random number, id is the tag's ID, k is a pre-shared secret
between the tag and the reader, and $\{f_n\}_{n \in \mathbb{N}}$ is a pseudo-random number function
ensemble. The reader exhaustively searches its database for a tag whose ID and key
can produce a match with the received data. The hash-lock protocol has a serious
efficiency problem that the reader needs to perform $O(n)$ hash computations on
line per authentication, where n is the number of tags in the system. Some variants
[1, 2, 14, 19] of hash-lock scheme try to improve the search efficiency, but they have
issues. The OKS protocol [14] uses hash-chain for anonymous authentication. The
OSK/AO protocol [1, 2] leverages the time-memory tradeoff to reduce the search
complexity to $O(n^{\frac{2}{3}})$ (still too large) at the cost of $O(n^{\frac{2}{3}})$ units of memory. However,
both OKS and OSK/AO cannot guarantee anonymity under denial-of-service (DoS)
attack [9]. The YA-TRAP protocol [19] makes use of monotonically increasing
timestamps to achieve anonymous authentication. YA-TRAP is also susceptible to
DoS attack, and a tag can only be authenticated once in each time unit. The DoS
attack in OSK/AO and YA-TRAP is in nature a desynchronization attack, which
tricks a tag into updating its keys unnecessarily and makes it fail to be authenticated
by an authorized reader later.

The LAST protocol was designed based on a weak privacy model [13]. Key
identifiers are used to facilitate the reader to identify the tags quickly. After each
authentication, the reader uploads a new $\langle identifier, key \rangle$ pair to the tag. LAST only
requires the reader and tag to compute $O(1)$ hashes per authentication, but the overhead for the reader to search a given key identifier is not considered. Moreover, since
the key identifier is only updated after a successful authentication, the tag keeps
sending the same key identifier between two consecutive successful authentications.
Therefore, LAST is not anonymous in the strict sense. In addition, the process of
uploading a new $\langle identifier, key \rangle$ pair to the tag after each authentication incurs extra
communication overhead.

3.2.2 Tree-Based Protocols

Tree-based protocols organize the shared keys in a balanced tree to reduce the complexity of identifying a tag to $O(\log n)$. However, the tree-based protocols generally require each tag to store $O(\log n)$ keys, which is $O(1)$ for non-tree-based protocols.

In Dimitriou's protocol [6], the non-leaf nodes of the tree store auxiliary keys that can be used to infer the path leading to leaf nodes that store the authentication keys. For each authentication, the computation overhead for both the reader and the tag is $O(\log n)$, and the tag needs to transit $O(\log n)$ hash values. This protocol is vulnerable to the compromising attack since different tags may share auxiliary keys [11, 12].

The ECNP protocol [10] leverages a cryptographic encoding technique to compress the authentication data transmitted by tags. ECNP can reduce the computation overhead of the reader and the transmission overhead of the tag by multifold compared with Dimitriou's protocol [6], but they remain $O(\log n)$ due to the use of tree structure. Moreover, ECNP is not resistant against the compromising attack since the children of one node in the tree share the same group keys.

The ACTION protocol [12] was designed to be resistant against the compromising attack. It adopts a sparse tree architecture to make the keys of each tag independent from one another. In ACTION, each tag is randomly assigned with a path key, which is further segmented into link indices to guide the reader to walk down the tree towards the leaf node that carries the secret key k of the tag. For each authentication, a tag needs to compute and transmit $O(\log n)$ hashes and the reader needs to perform $O(\log n)$ hashes to locate the shared key. The key problem of ACTION is that the size of link indices is too small after segmentation (e.g., 4 bits), rendering them easy to guess.

3.3 A Strawman Solution

We first introduce a strawman solution for lightweight anonymous authentication using pre-stored tokens.

3.3.1 Motivation

Most prior wok, if not all, employs cryptographic hash functions to randomize authentication data for the purpose of keeping anonymity. Implementing a typical cryptographic hash function such as MD4, MD5, and SHA-1 requires at least 7K logic gates [3]. However, widely used passive tags only have 7K–15K logic gates, of which 2K–5K are reserved for security purposes [16]. The hardware constraint necessitates a new design paradigm for lightweight anonymous authentication protocols that are more supportive for low-cost tags. The commercial success of RFID tags lies with their simplicity. Although there is no specification on how simple these tags should be, it is safe to say that we will always prefer a solution that achieves the comparable goal with less hardware requirement. On the other hand, the significant disparity between the readers and tags points out an asymmetry design principle that we will follow: push most complexity to the readers while leaving the tags as simple as possible.

3.3.2　A Strawman Solution

Consider an RFID system with n tags $t_1, t_2, \ldots, t_n$, each pre-installed with an array of m unique random tokens, $[tk_i^1, tk_i^2, \ldots, tk_i^m]$ $(1 \leq i \leq n)$. Tag t_i also has a token index pt_i (initialized to 1) pointing to its first unused token. The tokens and token index of each tag are also stored in the database of the central server, as illustrated in Table 3.2.

In the sequel, we consider the authentication process between an arbitrary reader R and an arbitrary tag t having a token array $[tk^1, tk^2, \ldots, tk^m]$ and a token index pt. To authenticate t, R sends a request to t. Upon receiving the request, t sends its first unused token tk^{pt} to R. After that, t increases pt by 1 to guarantee that the same token will not be used twice. Otherwise, t will always send the same token when an unauthorized reader requests a token, which breaks the anonymity. After receiving the token, R has to search the token in the database since it does not know the tag's identity. Starting from $i = 1$, R checks if $tk_i^{pt_i} = tk^{pt}$ one by one. If there exists an $i \in [1, n]$ such that $tk_i^{pt_i} = tk^{pt}$, t is successfully authenticated; otherwise, t fails the authentication. In the former case, R sends back the token $tk_i^{pt_i+1}$ to t to authenticate itself, and sets $pt_i = pt_i + 2$. Tag t compares the received token with tk^{pt} to authenticate R, and increases pt by 1 again. Figure 3.2 shows the three steps of the mutual authentication.

In this approach, the online computation overhead of the tag is low—only one comparison (requires far less hardware than implementing a cryptographic hash) per authentication. The online computation complexity of the reader is $O(n)$ since at most n comparisons (though one comparison is much cheaper than computing one hash value) are needed for searching the received token. The communication overhead for both the reader and the tag is $O(1)$.

Table 3.2 Key table for the preliminary design

Tag	Token array	Token index
t_1	$[tk_1^1, tk_1^2, \ldots, tk_1^m]$	pt_1
t_2	$[tk_2^1, tk_2^2, \ldots, tk_2^m]$	pt_2
$\vdots$	$\vdots$	$\vdots$
t_n	$[tk_n^1, tk_n^2, \ldots, tk_n^m]$	pt_n

Fig. 3.2 Three steps of token-based mutual authentication

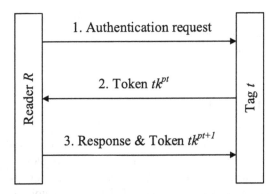

1. Authentication request

2. Token tk^{pt}

3. Response & Token tk^{pt+1}

Reader R

Tag t

To avoid the leakage of the tag's identity, the tokens used for authentication should look random. In addition, each token can be used only once. Hence, the tag must be replenished with new tokens after $\frac{m}{2}$ mutual authentications, e.g., purchasing new tokens from an authorized branch. Therefore, the tag should store as many tokens as possible to reduce the inconvenience caused by token replenishment. A low-cost tag, however, only has a tiny memory. For example, a passive UHF tag generally has a 512-bit user memory for storing user-specific data (tokens in our case). Some high-end tags with large memory [8, 15] are prohibitively expensive to be applied in large quantities. As an example, x Sky-ID tags [8] with an 8 KB user memory cost $25 each. We will introduce the security issues of this design in the next section.

3.4 Dynamic Token-Based Authentication Protocol

In this section, we describe our first dynamic Token-based Authentication Protocol (TAP). TAP can produce tokens for anonymous authentication on demand, and therefore does not require the tags to pre-install many tokens. However, we will show shortly that TAP still has some problems, which will be solved by our final design in the next section.

3.4.1 Motivation

Given the memory constraint, each tag can only store a few tokens. Frequent token replenishment brings about unacceptable inconvenience in practice. Hence, we want to invent a way to enable dynamic token generation from the few pre-installed tokens. In addition, the time for the reader to search a particular token is $O(n)$ in the preliminary design. We desire to reduce this overhead to $O(1)$. More importantly, we hope all advantages of the preliminary design, including no requirement of cryptographic hash functions, low computation overhead for the tag, and low communication overhead for both the reader and the tag, can be retained in our new design.

3.4.2 Overview

Let an arbitrary tag t in the system be pre-installed with u base tokens, denoted by $[bt^1, bt^2, \ldots, bt^u]$, each being a-bit long. These base tokens can be used to derive dynamic tokens for authentication. In addition, we introduce another type of auxiliary keys called base indicators to generate indicators that support the derivation of dynamic tokens. Suppose t stores v base indicators denoted by $[bi^1, bi^2, \ldots, bi^v]$, each being b-bit long. Let tk represent the current a-bit token, and ic be the current b-bit indicator. All the base tokens, base indicators, token, and indictor are also stored at the central server. Our idea is to let the reader and the tag independently generate the same random tokens by following the instruction encoded in the indicator. TAP consists of three phases: initialization phase, authentication phase, and updating phase, which will be elaborated one by one.

3.4.3 Initialization Phase

The central server stores all tags' keys in a key table, denoted by KT. As shown in Table 3.3, each entry is indexed by the tag index, supporting random access in $O(1)$ time. With the tag index idx, the keys of t can be found at $KT[idx]$.

When t joins the system, the reader randomly generates an array of u base tokens $[bt^1, bt^2, \ldots, bt^u]$, an array of v base indicators $[bi^1, bi^2, \ldots, bi^v]$, a token tk and an indicator ic for t. After that, the reader requests the central server to store those keys of t in the database. The central server inserts the keys to the first empty entry in KT. The search process for an empty entry can be sped up by maintaining a small table recording all empty entries in KT (e.g., due to tags' departure). If KT is fully occupied, the central server doubles its size to accommodate more tags.

Table 3.3 Key table stored by the central server for TAP

Tag index	Tag	Base token array	Token	Base indicator array	Indicator
1	t_1	$[bt_1^1, \ldots, bt_1^u]$	tk_1	$[bi_1^1, \ldots, bi_1^v]$	ic_1
2	t_2	$[bt_2^1, \ldots, bt_2^u]$	tk_2	$[bi_2^1, \ldots, bi_2^v]$	ic_2
$\vdots$	$\vdots$	$\vdots$	$\vdots$	$\vdots$	$\vdots$
n	t_n	$[bt_n^1, \ldots, bt_n^u]$	tk_n	$[bi_n^1, \ldots, bi_n^v]$	ic_n

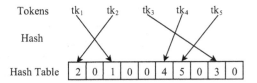

Fig. 3.3 A hash table used by TAP. The tokens of the five tags t_1, t_2, t_3, t_4, t_5 are $tk_1, tk_2, tk_3, tk_4, tk_5$, respectively. Each token is randomly mapped to a slot in the hash table, where the corresponding tag index is stored

To identify a tag based on its token in $O(1)$ time, the central server maintains a hash table HT, mapping the token of each tag to its tag index. Let HT consist of l slots. At first, every slot in HT is initialized to zero. After t joins the system, the reader computes the hash value $h(tk)$, where the hash function $h(\cdot)$ yields random values in $[1, l]$, and then puts the tag index idx of t in the $h(tk)$th slot of the hash table, i.e., $HT[h(tk)] = idx$ (the potential problem of hash collisions will be addressed shortly). Figure 3.3 presents an illustration of the hash table built for the tokens of five tags.

3.4.4 Authentication Phase

The authentication process of TAP is similar to that of the preliminary design as shown in Fig. 3.2. One difference is that the reader can quickly identify the tag from its token using the hash table. After receiving a token tk from tag t, the reader first calculates $h(tk)$, and then accesses $HT[h(tk)]$ to retrieve the tag index of t, which is idx. If the reader finds $idx = 0$, it asserts the tag is fake and informs the tag of authentication failure. Otherwise, the reader refers to $KT[idx]$ in the key table to fetch the token, and compares it with the received token tk. Only if the two tokens are identical will the tag pass the authentication. In case that t is successfully authenticated, the reader will generate and transmit a new token to authenticate itself. The generation of tokens with good randomness requires the reader (more exactly, the central server) and the tag to update their shared keys synchronously.

3.4.5 Updating Phase

To guarantee anonymity, the tokens exchanged between the reader and the tag should have good randomness. Therefore, the reader (central server) and the tag need to synchronously update their shared keys after the current token is used. We stress that the tag will update its keys once it uses its current token. Therefore, the same token will never be used in two consecutive authentications, which fundamentally differs from LAST [13] where the tag only updates its key identifier after a successful authentication (which breaks the anonymity).

Fig. 3.4 The structure of a
b-bit indicator

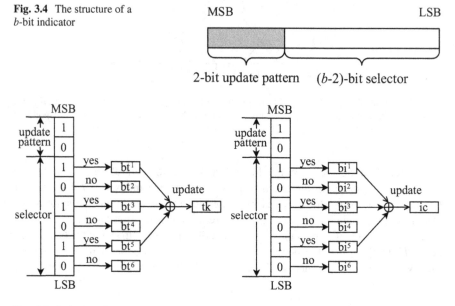

Fig. 3.5 *Left plot*: Generating a new token using the base tokens and the selector. *Right plot*: Generating a new indicator using the base indicators and the selector

The tag t relies on its current indicator ic to update its keys. Figure 3.4 shows the structure of an indicator, which includes two parts: The low-order $(b-2)$ bits form a selector, indicating which base tokens/base indicators should be used to derive the new token/indicator, while the high-order two bits encode the update pattern. When the updating phase begins, t calculates a new token from the base tokens according to the selector. Each of the low-order u bits ($u \leq b-2$) in the selector encodes a choice of the corresponding base token: "0" means not selected, while "1" means selected. For all selected base tokens, they are XORed to compute the new token. Therefore,

$$tk = \bigoplus_{j=1}^{u} ic[j]bt^j, \qquad (3.1)$$

where $ic[j]$ is the jth bit in ic (assume one-based indexes are used) and $\oplus$ is the XOR operator. The left plot in Fig. 3.5 gives an example of token update, where bt^1, bt^3, and bt^5 among the six base tokens happen to be selected. Similarly, t derives a new indicator from the base indicators as follows:

$$ic = \bigoplus_{j=1}^{v} ic[j]bt^j. \qquad (3.2)$$

At the server's side, the same new token and new indicator can be generated because it shares the same keys with the tag. In addition, the server also needs to update the hash table. First, the server sets $HT[h(tk)]$ (the old token) to 0, and after generating the new token, it sets $HT[h(tk)] = idx$.

After updating the token and the indicator, the central server and the tag need to further update the selected base tokens and base indicators. The update process for any selected base token or base indicator includes two steps: A one-bit left circular shift, and bit flip by following the particular 2-bit update pattern:

1. Pattern $(00)_2$: no flip is performed;
2. Pattern $(01)_2$: flip the jth bit if $j \equiv 0$ (mod 3);
3. Pattern $(10)_2$: flip the jth bit if $j \equiv 1$ (mod 3);
4. Pattern $(11)_2$: flip the jth bit if $j \equiv 2$ (mod 3).

Obviously, the ith and jth bits can be flipped together if and only if $i \equiv j$ (mod 3). This rationale of the updating scheme is that if the parameters a and b are set properly, any two bits in a base token or a base indicator have a chance to not be flipped together, thereby reducing their mutual dependence. We will provide the formal proof shortly. We emphasize that all keys are only stored at the central server rather than every single reader. Hence, the update process of a tag's keys triggered by one reader is transparent to other readers (a tag carrier can only appear at one location at a time).

3.4.6 Randomness Analysis

As required by our anonymous model, the tokens generated by TAP should be random and unpredictable. Randomness is a probabilistic property that should be described in terms of probability. We first prove the following theorem:

Theorem 1. *If $a \geq 2$ and $a \not\equiv 0$ (mod 3), there must exist a positive integer w, where $1 \leq w \leq a$, such that any two different bits in one base token will move to positions that cannot be flipped together after the base token is updated w times.*

Proof. We track two arbitrary bits in the base token bt^j, denoted by random variables X and $Y \in \{0, 1\}$. Suppose X and Y are initially located at the pth bit and qth bit of bt^j ($1 \leq p < q \leq a$), respectively, and w updates are performed ($1 \leq w \leq a$). Two possible cases need to be considered according to their initial positions:

Case 1: $q-p \not\equiv 0$ (mod 3) and $a+p-q \not\equiv 0$ (mod 3). First, if $q+w \leq a$, X and Y have moved to $bt^j[p+w]$ and $bt^j[q+w]$, respectively. Since $(q+w)-(p+w) \not\equiv 0$ (mod 3), they cannot be flipped together. Second, if $p + w \leq a < q + w$, then X moves to $bt^j[p+w]$ and Y moves to $bt^j[q+w-a]$. Because $(p+w)-(q+w-a) \not\equiv 0$ (mod 3), they still cannot be flipped together. Finally, if $p + w > a$, X and Y are now at $bt^j[p + w - a]$ and $bt^j[q + w - a]$, respectively. Similarly, since $(q + w - a) - (p+w-a) \not\equiv 0$ (mod 3), they cannot be flipped together. Hence, X and Y will never be flipped together under such conditions.

Case 2: $q - p \equiv 0 \pmod 3$ or $a + p - q \equiv 0 \pmod 3$ (note that by no means will $q - p \equiv a + p - q \equiv 0 \pmod 3$ because $a \not\equiv 0 \pmod 3$). If $q - p \equiv 0 \pmod 3$ and $a - q < w \le a - p$, X moves to $bt^j[p + w]$ and Y moves to $bt^j[q + w - a]$. Because $(p + u) - (q + u - a) \not\equiv 0 \pmod 3$, they move to positions that cannot flipped together. On the contrary, if $a + p - q \equiv 0 \pmod 3$, X and Y will not be flipped together at the beginning, but it becomes possible after w updates as long as $a - q < w \le a - p$.

Hence, X and Y have a chance to not be flipped together within a updates regardless of their initial positions.

Before investigating the randomness of the derived tokens, we first study the randomness of an arbitrary base token during its updates. We have the following lemma:

Lemma 1. *If the update pattern in the indicator is random, an arbitrary bit in a base token becomes 0 or 1 with equal probability using our update scheme.*

Proof. Let us track one arbitrary bit in bt^j, denoted by a random variable $X \in \{0, 1\}$. Suppose X is currently located at position $bt^j[i]$, where $1 \le i \le a$. When bt^j is updated, X is left shifted and then flipped with a probability of 0.25 if the update pattern is random. Therefore, the transition matrix for X during

each update is $P_1 = \begin{pmatrix} \frac{3}{4} & \frac{1}{4} \\ \frac{1}{4} & \frac{3}{4} \end{pmatrix}$. Using singular value decomposition (SVD) [18],

$P_1 = \begin{pmatrix} \frac{\sqrt{2}}{2} & \frac{\sqrt{2}}{2} \\ \frac{\sqrt{2}}{2} & -\frac{\sqrt{2}}{2} \end{pmatrix} \begin{pmatrix} 1 & 0 \\ 0 & \frac{1}{2} \end{pmatrix} \begin{pmatrix} \frac{\sqrt{2}}{2} & \frac{\sqrt{2}}{2} \\ \frac{\sqrt{2}}{2} & -\frac{\sqrt{2}}{2} \end{pmatrix}$. Hence, the transition matrix for X after

w updates is

$$P_1{}^w = \begin{pmatrix} \frac{\sqrt{2}}{2} & \frac{\sqrt{2}}{2} \\ \frac{\sqrt{2}}{2} & -\frac{\sqrt{2}}{2} \end{pmatrix} \begin{pmatrix} 1 & 0 \\ 0 & \frac{1}{2} \end{pmatrix}^w \begin{pmatrix} \frac{\sqrt{2}}{2} & \frac{\sqrt{2}}{2} \\ \frac{\sqrt{2}}{2} & -\frac{\sqrt{2}}{2} \end{pmatrix} = \begin{pmatrix} \frac{1}{2} + \frac{1}{2}^{w+1} & \frac{1}{2} - \frac{1}{2}^{w+1} \\ \frac{1}{2} - \frac{1}{2}^{w+1} & \frac{1}{2} + \frac{1}{2}^{w+1} \end{pmatrix},$$

which converges to $\begin{pmatrix} \frac{1}{2} & \frac{1}{2} \\ \frac{1}{2} & \frac{1}{2} \end{pmatrix}$. Therefore, X becomes 0 or 1 with equal probability.

Now let us further investigate two arbitrary bits in a base token, and we have the following lemma:

Lemma 2. *If the update pattern in the indicator is random, two arbitrary bits in a base token are independent under our update scheme.*

Proof. Consider two arbitrary bits, denoted by random variables X and Y, in base token bt^j. Suppose X and Y are initially located at the pth bit and qth bit of bt^j ($1 \le p < q \le a$), respectively. The transition matrices when X and Y cannot be flipped together and can be flipped together are

$$P_2 = \begin{pmatrix} \frac{1}{2} & \frac{1}{4} & \frac{1}{4} & 0 \\ \frac{1}{4} & \frac{1}{2} & 0 & \frac{1}{4} \\ \frac{1}{4} & 0 & \frac{1}{2} & \frac{1}{4} \\ 0 & \frac{1}{4} & \frac{1}{4} & \frac{1}{2} \end{pmatrix}, \text{ and } P_3 = \begin{pmatrix} \frac{1}{2} & 0 & 0 & \frac{1}{2} \\ 0 & \frac{1}{2} & \frac{1}{2} & 0 \\ 0 & \frac{1}{2} & \frac{1}{2} & 0 \\ \frac{1}{2} & 0 & 0 & \frac{1}{2} \end{pmatrix},$$

respectively. Assume that among the w ($w \geq a$) updates, X and Y cannot be flipped together for β times, while can be flipped together for γ times. We know from Theorem 1 that $\beta \geq 1$, so we have $\beta \geq 1$, $\gamma \geq 0$, and $\beta + \gamma = w$.

Case 1: $\gamma > 0$. We have

$$P_2^{\beta} \times P_3^{\gamma} = \begin{pmatrix} \frac{1}{4} & \frac{1}{4} & \frac{1}{4} & \frac{1}{4} \\ \frac{1}{4} & \frac{1}{4} & \frac{1}{4} & \frac{1}{4} \\ \frac{1}{4} & \frac{1}{4} & \frac{1}{4} & \frac{1}{4} \\ \frac{1}{4} & \frac{1}{4} & \frac{1}{4} & \frac{1}{4} \end{pmatrix}$$

for any combinations of β and γ.

Case 2: $\gamma = 0$. Hence, $P_2^{\beta} \times P_3^{\gamma} = P_2^{w}$. Using SVD, we can calculate

$$P_2^{w} = \begin{pmatrix} \frac{1}{4} + \frac{1}{2}^{w+1} & \frac{1}{4} & \frac{1}{4} & \frac{1}{4} - \frac{1}{2}^{w+1} \\ \frac{1}{4} & \frac{1}{4} + \frac{1}{2}^{w+1} & \frac{1}{4} - \frac{1}{2}^{w+1} & \frac{1}{4} \\ \frac{1}{4} & \frac{1}{4} - \frac{1}{2}^{w+1} & \frac{1}{4} + \frac{1}{2}^{w+1} & \frac{1}{4} \\ \frac{1}{4} - \frac{1}{2}^{w+1} & \frac{1}{4} & \frac{1}{4} & \frac{1}{4} + \frac{1}{2}^{w+1} \end{pmatrix},$$

each entry converging to $\frac{1}{4}$. Therefore, two arbitrary bits in a base token are pairwise independent.

With the two lemmas above, we can prove the following theorem regarding the randomness of the derived tokens:

Theorem 2. *If the indicator is random, any bit in the derived token has an equal probability to be 1 or 0, and two arbitrary bits in the derived token are independent using our update scheme.*

Proof. Consider the ith bit of the derived token tk, denoted by $tk[i]$ ($1 \leq i \leq a$). We know $tk[i] = \bigoplus_{j=1}^{u} ic[j] bt^j[i]$. Let N_0 be the random variable of the number of base tokens whose ith bit is 0, and N_1 be the random variable of the number of base tokens whose ith bit is 1, subjecting to $N_0 \geq 0$, $N_1 \geq 0$ and $N_0 + N_1 = u$. According to Lemma 1 and the independence of different base tokens, N_0 follows the binomial distribution $B(u, 0.5)$, and $P(N_0 = x) = \binom{u}{x} \times \left(\frac{1}{2}\right)^u$, where $0 \leq x \leq u$. To calculate $tk[i]$, we need to consider two possible cases:

Case 1: $N_0 = u$, namely there is no 1 in those u bits. In this case, $tk[i]$ must be 0.

Case 2: $0 \leq N_0 < u$. In this case, $tk[i]$ can be 0 or 1. If $tk[i] = 0$, it implies that an even number of base tokens whose ith bit is 1 are chosen, and the conditional probability is

$$P(tk[i] = 0 \,|\, 0 \le N_0 < u) = \frac{2^{N_0} \times \sum_{x=0}^{\lceil \frac{N_1}{2} \rceil} \binom{N_1}{2x} - 1}{2^u - 1}$$ (3.3)

$$= \frac{2^{u-1} - 1}{2^u - 1}.$$

Hence, the probability for $tk[i] = 0$ is

$$\begin{aligned}
P(tk[i] = 0) &= P(N_0 = u) \times P(tk[i] = 0 \,|\, N_0 = u) \\
&\quad + P(0 \le N_0 < u) \times P(tk[i] = 0 \,|\, 0 \le N_0 < u) \\
&= \frac{1}{2^u} \times 1 + (1 - \frac{1}{2^u}) \times \frac{2^{u-1} - 1}{2^u - 1} = \frac{1}{2}.
\end{aligned}$$ (3.4)

Therefore, $P(tk[i] = 1) = P(tk[i] = 0) = \frac{1}{2}$. Moreover, because $tk[i]$ is determined only by the ith bits of the base tokens, and two arbitrary bits in a base tokens are independent according to Lemma 2, two arbitrary bits in the derived token are also independent.

The randomness analysis of the indicators follows the same path. The simulation results provided in Sect. 3.7 demonstrate that the tokens and indicators have very good randomness.

3.4.7 Discussion

Memory Requirement To implement TAP, each tag needs $(u + 1)a + (v + 1)b$ bits of memory to store the keys. Our simulation results in Sect. 3.7 show that a, b, u, and v can be set as small constants. Therefore, the memory requirement for the tag is small. The memory requirement at the central server is $O(n)$ for storing the key table and hash table.

Communication Overhead For each authentication, the tag only needs to transmit one a-bit token, and the reader sends an authentication request and one a response, both incurring $O(1)$ communication overheads.

Online Computation Overhead For each authentication, the tag generates two tokens and performs one comparison to authenticate the reader. All operations performed by the tag, including bit-wise XOR, bit flip, and one-bit left circular shift, are simple and hardware efficient. The reader (or the server) needs to calculate two extra hash values: one for the token received from the tag to identify the tag, and the other for the new token to update the hash table. Both the tag and the reader have $O(1)$ computation overhead.

3.4.8 Potential Problems of TAP

TAP has three potential problems that should be addressed.

Desynchronization Attack An unauthorized reader can also initiate an authentication by sending a request. The tag will reply with its current token, and then update its keys as usual. As a result, its keys differ from what are stored by the central server. When the tag encounters a legitimate reader later, it will probably fail the authentication as its current token does not match the one stored in the central server.

Replay Attack When performing a desynchronization attack, the adversary can record the received token. Later it can retransmit the token to authenticate itself. Since the token is valid, it will pass the authentication. The above two issues also exist in the preliminary design.

Hash Collision For two tags in the system, the hash values of their current tokens may happen to be the same, called hash collision. In this case, their tag indexes cannot be stored in the same slot of the hash table. Otherwise, the reader cannot uniquely identify the tag through the received token. In addition, since each tag generates its tokens independently, it may happen that two tags have the same token, called token collision. Token collision is a special case of hash collision, and token collision must lead to hash collision. We find that hash collisions, though the probability is low, can cause problems to all anonymous RFID authentication protocols using cryptographic hashes, but the potential problems are never carefully addressed.

3.5 Enhanced Dynamic Token-Based Authentication Protocol

In this section, we present our third protocol, called Enhanced dynamic Token-based Authentication Protocol (ETAP), to address the issues of TAP.

3.5.1 Resistance Against Desynchronization and Replay Attacks

Since desynchronization attack and replay attack can be carried out simultaneously, we tackle them together. Our objective is twofold: First, the valid tag can still be successfully authenticated by a legitimate reader after some desynchronization attacks; Second, even if the adversary has captured some tokens from the valid tag, it cannot use those tokens to authenticate itself.

Table 3.4 Key table stored by the central server for ETAP

Tag index	Tag	Base token array	Token array	Base indicator array	Indicator
1	t_1	$[bt_1^1, \ldots, bt_1^u]$	$[tk_1^1, tk_1^2, \ldots, tk_1^k]$	$[bi_1^1, \ldots, bi_1^v]$	ic_1
2	t_2	$[bt_2^1, \ldots, bt_2^u]$	$[tk_2^1, tk_2^2, \ldots, tk_2^k]$	$[bi_2^1, \ldots, bi_2^v]$	ic_2
$\vdots$	$\vdots$	$\vdots$	$\vdots$	$\vdots$	$\vdots$
n	t_n	$[bt_n^1, \ldots, bt_n^u]$	$[tk_n^1, tk_n^2, \ldots, tk_n^k]$	$[bi_n^1, \ldots, bi_n^v]$	ic_n

Fig. 3.6 Our scheme against desynchronization attack

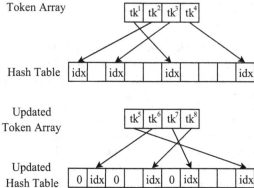

To make our protocol resistant against desynchronization attack, we let the central server pre-calculate an array of k tokens $[tk^1, tk^2, \ldots, tk^k]$ from the base tokens, and any token can be used to identify the tag, where k is a system parameter that can be set large or small, depending on the available memory at the server. The reader needs at least one token to identify the tag, and thus at most $k-1$ desynchronization attacks can be tolerated.[1] After a successful mutual authentication, the reader will replenish the token array with k new tokens. Furthermore, we use a two-step verification process to guard against replay attacks. Table 3.4 shows the key table stored by the central server for implementing ETAP.

Now let us elaborate ETAP with an example given in Fig. 3.6. Suppose $k = 4$ and the reader pre-calculates four tokens tk^1, tk^2, tk^3, and tk^4 for tag t with tag index idx. In addition, suppose the current token stored by t is $tk = tk^2$, which means t may have been under one desynchronization attack and the adversary has captured tk^1. When the reader receives tk^2 from t, it accesses $HT[h(tk^2)]$ to fetch the tag index idx of the t, and then the token array of t from $KT[idx]$. The reader then traverses the token array until it finds tk^2. After that, the reader uses the next token in the token array, tk^3 in this example, to authenticate itself. If the received token happens to be at tail of the array, the reader needs to derive a new token for authentication. To prevent the adversary from passing authentication using tk^1,

---

[1]An exponentially increasing timeout period can be enforced between unsuccessful authentications to prevent an adversary from depleting the k tokens too quickly.

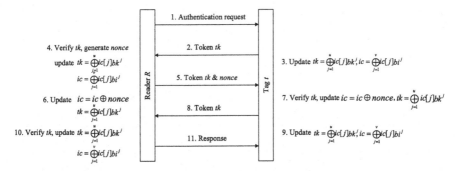

Fig. 3.7 Two-step verification mechanism of ETAP

we adopt the two-step verification as illustrated in Fig. 3.7. In step 3, the reader includes a b-bit random nonce in its message, and challenges the tag to send another token. After the tag authenticates the reader, it updates its indicator by XORing the indicator with the received nonce (so does the reader), which contributes to randomizing the indicator as well. After that, the tag derives a new token based on the updated indicator, and sends it to the reader for the second verification. Since the adversary does not know the base tokens and the indicator, it cannot derive the correct token to pass the second verification, rendering replay attack infeasible. After the successful mutual authentication, the reader generates four new tokens to replenish the token array. In addition, the reader updates HT by setting the slots corresponding to the old tokens to 0, and setting the slots corresponding to the new tokens to idx. Note that the token replenishment is performed off line by the central server, which is therefore not a performance concern.

3.5.2 Resolving Hash Collisions

Suppose the central server pre-computes k tokens for each tag. Let $n_t = nk$ be the number of total tokens, and l be the size of the hash table. A slot in the hash table is called an *empty slot*, a *singleton slot*, and a *collision slot*, respectively, if zero, one and multiple tokens are mapped to it. When every token is mapped to a singleton slot (no collision happens), the *utilization ratio* ρ of the hash table is defined as

$$\rho = \frac{nk}{l} = \frac{n_t}{l}, \tag{3.5}$$

which is used as the performance metric for evaluating memory efficiency.

One candidate approach for reducing hash collisions is to use a large hash table. However, l must be set prohibitory large for the purpose of totally eliminating hash collisions, resulting in low memory utilization (small ρ). Figure 3.8 shows the memory utilization ratio of the hash table when different numbers of tokens need to

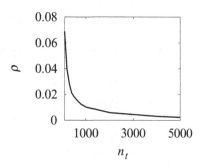

Fig. 3.8 Utilization ratio of the hash table when a single hash function is used

be stored and a single hash function is used. We can see the utilization ratio ρ is very low. Moreover, the value of ρ drops dramatically when more tokens are mapped to the hash table. For example, $\rho = 0.0023$ when $n_t = 5000$, meaning more than 99 % of slots in the hash table are wasted.

We observe that two different tokens causing a hash collision under one hash function probably will not have a collision under another hash function. Therefore, using multiple hash functions provides an alternative way for resolving hash collisions.

When a single hash function is used, the probability p_s that an arbitrary slot is a singleton slot is

$$
\begin{aligned}
p_s &= \binom{n_t}{1} \frac{1}{l} \left(1 - \frac{1}{l}\right)^{n_t - 1} \\
&\approx \frac{n_t}{l} e^{-\frac{n_t - 1}{l}} \\
&\approx \frac{n_t}{l} e^{-\frac{n_t}{l}}.
\end{aligned}
\tag{3.6}
$$

It is easy to prove that $p_s \leq \frac{1}{e} \approx 0.368$ and it is maximized when $l = n_t$. In contrast, if we apply two independent hash functions to map tokens to slots, a slot will have a probability of up to $1 - (1 - 0.368)^2 \approx 0.601$ to be a singleton in one of the two mappings. Similarly, if we apply r independent hash functions from tokens to slots, the probability that a slot will be a singleton in one of the r mappings can increase to $1 - (1 - 0.368)^r$, which quickly approaches to 1 with the increase of r. Figure 3.9 presents an example showing the advantage of using multiple hash functions in reducing hash collisions. In the left plot, only one hash function is used, and there is only one singleton slot, while in the right plot, three hash functions are employed and every token is mapped to a singleton slot.

Inspired by the above observation, ETAP employs r independent hash functions, denoted by $h_1, h_2, \ldots, h_r$, to resolve hash collisions and improve memory efficiency. Every hash function can generate hash values uniformly distributed over $[1, l]$. To insert a token tk of tag t to the hash table, the reader calculates $h_i(tk)$ in sequence

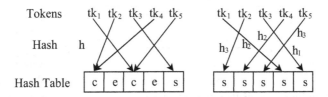

Fig. 3.9 An example of using multiple hash functions to reduce hash collisions, where the *left plot* uses one hash function, and the *right plot* uses three hash functions. h, h_1, h_2, and h_3 are hash functions. "e" means an empty slot, "s" means a singleton slot, and "c" means a collision slot

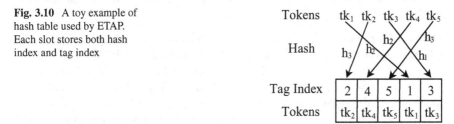

Fig. 3.10 A toy example of hash table used by ETAP. Each slot stores both hash index and tag index

starting from $i = 1$. If the $h_i(tk)$th slot in the hash table has not been occupied by any token, tk can be added to this slot immediately. Each slot needs to store both the token value and the tag index associated with the token to facilitate identification of which token indeed uses a certain slot. As the example in Fig. 3.10, a token tk_2 of t_2 is mapped by the hash function $h_3(\cdot)$ to the first slot (a singleton slot) of HT. Hence, $HT[0]$ records tag index 2 and the token value tk_2. Instead of implementing r independent hash functions, we can use one master hash function H and a set S of random seeds, and let

$$h_i(tk) = H(tk \oplus S[i]), \tag{3.7}$$

where $\oplus$ is the XOR operator. When the reader receives a token tk from a tag for authentication, it computes $h_i(tk)$ ($1 \le i \le r$) until it finds that the token value in slot $HT[h_i(tk)]$ matches tk, where it can obtain the correct tag index of that tag. If no matching token is found, the tag fails authentication.

We will shortly evaluate the effectiveness of our scheme for resolving hash collisions caused by different tokens through simulations. The issue of token collisions, however, may not be solved by this approach. If the two identical tokens happen to be associated with the same tag, it will not cause any problem. But if they are associated with different tags, the reader cannot uniquely identify the tag from the received token. Therefore, such collided tokens cannot be used for authentication. The central server can store those tokens in a CAM (Content Addressable Memory) [5] or a Bloom filter for quick lookup. When the reader receives a token, it first checks if it will cause token collision; if so, the reader needs to request another token from the tag for identification purpose. We expect that the chance for token collisions is small as long as the generated tokens have good randomness.

3.5.3 Discussion

Memory Requirement The memory requirement for the tag to implement ETAP is the same as TAP, i.e., $(u + 1)a + (v + 1)b$ bits. The memory requirement at the central server moderately increases because of the larger key table and hash table for storing multiple tokens for each tag.

Communication Overhead For each authentication, the tag only needs to transmit two a-bit tokens, and the reader needs to send an authentication request, one a-bit token, one b-bit nonce, and a response, both incurring $O(1)$ communication overheads.

Online Computation Overhead For each authentication, the tag generates three tokens and performs one comparison to authenticate the reader. ETAP requires some extra computation overhead from the reader (server). First, the reader should check if a received token is a collided one, which requires $O(1)$ computation. In addition, the reader needs to calculate at most r hash values to identify the tag, and perform at most k comparisons to locate the received token in the token array. Since r and k are small constants, the online computation overhead for the reader is still $O(1)$.

Hardware Cost The hardware for RFID tags to implement ETAP consists of a circular shift register, two registers for storing intermediate results, some XOR gates, and some RAM to store u base tokens, v base indicators, one token, and one indicator. We estimate the hardware cost of ETAP following the estimated costs of typical cryptographic hardware [4, 16] listed in Table 3.5. The circular shift register is a group of flip-flops connected in chain, which requires $12 \times \max(a, b)$ logic gates. Similarly, the two registers for intermediate results need $2 \times 12 \times \max(a, b)$ logic gates. In addition, it takes $2.5 \times \max(a, b)$ logic gates to implement the XOR gates. Finally, the cost of the RAM for storing the base tokens, base indicators, token, and indicator is about $\frac{(u+1)a}{8} \times 12 + \frac{(v+1)b}{8} \times 12$ logic gates. Therefore, the total number of required logic gates for implementing ETAP is approximately $38.5 \times \max(a, b) + 1.5 \times (u+1)a + 1.5 \times (v+1)b$. For example, if we set $a = b = 16$, $u = 10$, and $v = 6$ (the reason for this setting will be explained shortly), ETAP only requires about 1K logic gates.

Table 3.5 Estimated costs of typical cryptographic hardware

Functional block	Cost (logic gates)
2 input NAND gate	1
2 input XOR gate	2.5
2 input AND gate	2.5
FF (Flip-flop)	12
n-byte RAM	$n \times 12$

3.6 Security Analysis

ETAP is designed to be resistant against desynchronization attack and replay attack. In this section, we further analyze the security of ETAP under both passive and active attacks.

Known Token Attack In ETAP, the tokens are transmitted without any protection, which may lead to a potential security loophole. The adversary can capture all tokens exchanged between the reader and the tag, and use them to infer the base tokens. However, we have the following theorem:

Theorem 3. *Cracking the base tokens from the captured tokens is computationally intractable if a sufficient number of base tokens are used.*

Proof. According to (3.1), each captured token provides an equation of the base tokens. Since there are u base tokens, at least u independent equations are needed to obtain a solution of the base tokens. However, the adversary has no clue about the current value of the indicator, which can have very good randomness as shown in Sect. 3.7. Therefore, the adversary has no better way than trying each possible value of the indicator by brute force. Hence, the u bits in the selector and the 2-bit update pattern give 2^{u+2} instantiations of each equation. Therefore, the adversary has to solve $(2^{u+2})^u = 2^{u(u+2)}$ different equation sets. For each candidate solution, the adversary derives another token, and compares it with the captured one to verify if the solution is correct, which requires another 2^u trials. As a result, the total computation overhead for the adversary to crack the base tokens is $2^u \times 2^{u(u+2)} = 2^{u(u+3)}$, which is computationally intractable if u is set reasonably large, e.g., $u = 10$.

Anonymity Due to the randomness of the tokens (verified in Sect. 3.7), the adversary cannot associate any tokens with a certain tag. According to Theorem 2, the probability that the adversary can successfully guess any bit z of a tag's next token based on its previous tokens is

$$Prob(z\prime = z) \leq \frac{1}{2} + \frac{1}{ploy(s)}, \tag{3.8}$$

where $z\prime$ is the adversary's guess of z, and $poly(s)$ is an arbitrary polynomial with a security parameter s. Therefore, the adversary does not have a non-negligible advantage in guess z, and ETAP can preserve the anonymity of tags.

Compromising Resistance In ETAP, the keys of each tag are initialized and updated independently. Even if all tags except two are compromised by an adversary, it still cannot infer the keys of the two remaining tags or distinguish them based on their transmitted tokens. Therefore, ETAP is robust against compromising attack.

Forward Secrecy Forward secrecy requires that an adversary cannot crack the previous messages sent by a tag even if the adversary obtains the current keys of the tag. ETAP has perfect forward secrecy because in step 7 of each authentication, the tag will XOR its current indicator with a random nonce. Even if the adversary obtains all current keys of the tag, it does not know the previous values of the indicator without capturing all random nonces. Therefore, the adversary cannot perform reverse operations of the updating process to calculate the previous tokens.

3.7 Numerical Results

In this section, we first evaluate the effectiveness of the multi-hash scheme in resolving hash collisions and improving memory efficiency as well. After that, we run randomness tests on the tokens generated by ETAP.

3.7.1 Effectiveness of Multi-Hash Scheme

In the first set of simulations, the number n_t of tokens is set to 100, 500, 1000, and 5000, respectively. We vary the number r of hash functions from 1 to 20. Under each parameter setting, we repeat the simulation 500 times and obtain the average value of utilization ratio. Results in Fig. 3.11 demonstrate that ρ increases significantly with the increase of r at first, and gradually flattens out when r is sufficiently large. Consider the case that $n_t = 5000$. When $r = 1$, less than 0.3 % of slots are used. In contrast, when $r = 10$, more than 50 % of slots are occupied. In addition, we observe that for larger n_t, the corresponding ρ is slightly smaller when the same number of hash functions is used.

Next, we investigate the effectiveness of the multi-hash approach in resolving hash collisions. We fix the number r of hash functions to 10, and vary the number n_t of tokens from 1000 to 10,000 at steps of 1000. The number l of slots in the hash table is set to 2×, 3×, and 5× the number of tokens, respectively. Under each parameter setting, we run 500 tests and calculate the ratio of tests that no hash collision occurs. The results in Fig. 3.12 show that when $l = 2n_t$, hash collisions can occur with high probability, particularly for large n_t; when l is increased to $5n_t$, there is no hash collision any more.

Fig. 3.11 Memory **utilization ratio** of the hash time when multiple hash functions are used

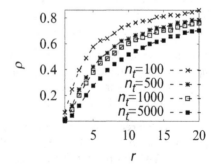

Fig. 3.12 Ratio of tests that
have no hash collision when
10 hash functions are used
($r = 10$)

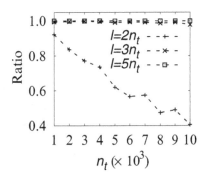

3.7.2 Token-Level Randomness

The effectiveness of ETAP relies on the randomness of the tokens and indicators.
An intuitive requirement of randomness is that any token (indicator) should have
approximately the same probability to appear. The EPC C1G2 standard [7] specifies
that for a 16-bit pseudo-random generator the probability of any 16-bit $RN16$ with
value x shall be bounded by $\frac{0.8}{2^{16}} < P(RN16 = x) < \frac{1.25}{2^{16}}$. To evaluate the randomness
of tokens and indicators generated by ETAP, we set $a = b = 16$, respectively,
produce $2^{16} \times 500$ tokens and indicators, and calculate the frequency of each token
or indicator. Note that we can concatenate multiple tokens to form a longer one if
necessary. In addition, we set $u = 10$ as suggested by Theorem 3, and vary $v = 2, 4,$
6 to investigate its impact on randomness. Figure 3.13 presents the results, where the
dotted horizontal lines represent the bounds specified by EPC C1G2. We can see that
the indicators have better randomness with the increase of v, while the randomness
of tokens is not sensitive to the value of v since u is already set sufficiently large.
In addition, when $u = 10$ and $v = 4$, requiring only 256-bit tag memory, both the
tokens and indicators meet the randomness requirement.

3.7.3 Bit-Level Randomness

The National Institute of Standards and Technology (NIST) provides a statistical
suite for randomness test [17], including monobit frequency test, block frequency
test, cumulative sums test, runs test, test for the longest run of ones in a block, binary
matrix rank test, etc. Due to space limitation, we cannot explain each test here, and
interested readers can refer to [17] for detail information. Given a sequence of n_s
bits, it is accepted as random only if the observed p-value is no less than a pre-
specified level of significance α based on the null hypothesis H_0.

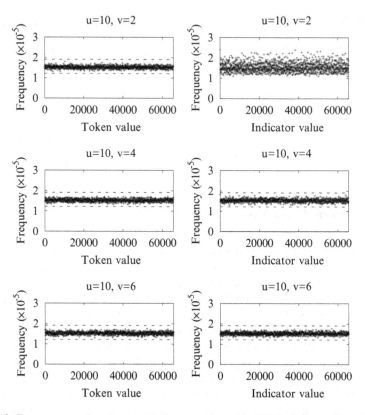

Fig. 3.13 Frequency tests for tokens and indicators generated by ETAP, where $a = b = 16$. Each point represents a token/indicator and its frequency. The two *dotted horizontal lines* represent the required bounds

We use two metrics to evaluate the test results: (1) The proportion of sequences that pass the tests. The acceptable range is $\hat{p} \pm 3\sqrt{\frac{\hat{p}(1-\hat{p})}{m_s}}$ [17], where $\hat{p} = 1 - \alpha$ and m_s is the sample size; (2) The uniformity of the observed p-values. Let X be a random variable with probability density function $f_X(x)$, and $Y \in [0, 1]$ be the p-value of X. Since the cumulative distribution function $F(X)$ of X is monotonically increasing, we have

$$P(Y \leq y) = P(\int_X^\infty f_X(x)dx \leq y) = P(1 - F(X) \leq y)$$

$$= 1 - P(X \leq F^{-1}(1-y)) = 1 - (1-y) = y. \tag{3.9}$$

Hence, $Y \sim U(0, 1)$. We divide $(0, 1)$ into ten equal-length subintervals, and denote the numbers of p-values in each subinterval as $F_1, F_2, ..., F_{10}$, respectively. We have

$$\chi^2 = \sum_{i=1}^{10} \frac{(F_i - \frac{m_s}{10})^2}{\frac{9m_s}{100}} \sim \chi^2(9). \tag{3.10}$$

The proof is given below:

Proof. Consider the value of F_i, where $1 \leq i \leq 10$. Let Z_{ij} be the event that the p-value of the jth ($1 \leq j \leq m_s$) sequence, denoted by Y_j, belongs to the ith subinterval $[\frac{i-1}{10}, \frac{i}{10})$. In addition, let $1_{Z_{ij}}$ be the corresponding indicator random variable, namely

$$1_{Z_{ij}} = \begin{cases} 1, & \text{if } Y_j \in [\frac{i-1}{10}, \frac{i}{10}), \\ 0, & \text{otherwise.} \end{cases}$$

Therefore, we have $F_i = \sum_{j=1}^{m_s} 1_{Z_{ij}}$. Since $Y_j \sim U(0, 1)$, we have $E(1_{Z_{ij}}) = \frac{1}{10}$ and $Var(1_{Z_{ij}}) = \frac{1}{10} \times (1 - \frac{1}{10}) = \frac{9}{100}$. Hence, $E(F_i) = \frac{m_s}{10}$, and $Var(F_i) = \frac{9m_s}{100}$. Based on the Central Limit Theorem (CLT), $\frac{F_i}{m_s}$ converges to $Norm(\frac{1}{10}, \frac{9}{100m_s})$ asymptotically. Therefore, $(\frac{\frac{F_i}{m_s} - \frac{1}{10}}{\frac{3}{10\sqrt{m_s}}})^2 \sim \chi^2(1)$, and $\chi^2 = \sum_{i=1}^{10}(\frac{\frac{F_i}{m_s} - \frac{1}{10}}{\frac{3}{10\sqrt{m_s}}})^2 = \sum_{i=1}^{10} \frac{(F_i - \frac{m_s}{10})^2}{\frac{9m_s}{100}} \sim \chi^2(9)$.

Therefore, we can employ χ^2 test. If the observed statistic of χ^2 is $\chi^2(obs)$, the p-value is

$$P(\chi^2 \geq \chi^2_{obs}) = \frac{\int_{\chi^2(obs)}^{\infty} e^{-x/2} x^{9/2-1} dx}{\Gamma(9/2) 2^{9/2}}$$

$$= \frac{\int_{\chi^2(obs)/2}^{\infty} e^{-x} x^{9/2-1} dx}{\Gamma(9/2)}$$

$$= igamc(\frac{9}{2}, \frac{\chi^2(obs)}{2}),$$

where $igamc(c, z) = 1 - \frac{\int_{-\infty}^{z} e^{-x} x^{c-1} dx}{\Gamma(c)}$. The uniformity is acceptable if $igamc(\frac{9}{2}, \frac{\chi^2(obs)}{2}) \geq 0.0001$ [17].

We set $a = b = 16$, $u = 10$, and $v = 4$, and convert the tokens generated by ETAP to a bit sequence. We vary n_s from 1000, 5000, 10,000 to 50,000. The NIST suggests that $\alpha \geq 0.001$, so we set $\alpha = 0.01$. In addition, we set $m_s = 500$, in the same order of magnitude as α^{-1}. The block size M should be selected such that $M \geq 20$, $M > 0.01n_s$, and $N_B < 100$, where N_B is the number of blocks. We set $M = 0.02n_s$, so $N_B = \frac{n_s}{M} = 50$.

The test results are shown in Table 3.6. We can see that the bit sequence generated by ETAP can pass the randomness tests under all parameter settings, which again verifies that our protocol can generate tokens with good randomness.

Table 3.6 The sample size m_s is 500, and the acceptable confidence interval of the success proportion is $[0.97665, 1]$

Test	Length 1000		5000		10,000		50,000	
	Prop.	p-value	Prop.	p-value	Prop.	p-value	Prop.	p-value
Monobit frequency	0.9920	0.002927	0.9880	0.861264	0.9920	0.719747	0.9900	0.957612
Block frequency	0.9960	0.037076	0.9880	0.538182	0.9880	0.162606	0.9940	0.055361
Cumulative sum (Mode 0)	0.9920	0.119508	0.9860	0.123755	0.9880	0.632955	0.9880	0.986227
Cumulative sum (Mode 1)	0.9920	0.798139	0.9920	0.823725	0.9900	0.877083	0.9900	0.081510
Runs	0.9940	0.286836	0.9820	0.482707	0.9880	0.146982	0.9860	0.068571
Longest run	0.9960	0.583145	0.9880	0.554420	0.9860	0.851383	0.9900	0.889118
Matrix rank[a]	NA	NA	NA	NA	NA	NA	0.9880	0.004697

[a]Matrix rank test requires that the bit sequence consists of at least 38,912 bits. Hence, no test is performed when $n_s < 38,912$, which is marked as NA

3.8 Summary

This chapter covers lightweight anonymous authentication in RFID systems. To meet the hardware constraint of low-cost tags, we abandon hardware-intensive cryptographic hashes and follow the asymmetry design principle. Our protocol ETAP uses a novel technique to generate random tokens on demand for anonymous authentication. The randomness analysis and tests demonstrate that ETAP can produce tokens with very good randomness. Moreover, ETAP reduces the communication overhead and online computation overhead to $O(1)$ per authentication for both the tags and the readers, which compares favorably with the prior art.

References

1. Avoine, G., Oechslin, P.: A scalable and provably secure hash-based RFID protocol. In: IEEE PerCom Workshops, pp. 110–114 (2005)
2. Avoine, G., Dysli, E., Oechslin, P.: Reducing time complexity in RFID systems. In: Selected Areas in Cryptography, pp. 291–306. Springer, Berlin/Heidelberg (2006)
3. Bogdanov, A., Leander, G., Paar, C., Poschmann, A., Robshaw, M.J.B., Seurin, Y.: Hash functions and RFID tags: mind the gap. In: Proceedings of CHES, pp. 283–299 (2008)
4. Chen, M., Chen, S., Xiao, Q.: Pandaka: a lightweight cipher for RFID systems. In: Proceedings of IEEE INFOCOM, pp. 172–180 (2014)
5. Chisvin, L., Duckworth, R.J.: Content-addressable and associative memory: alternatives to the ubiquitous RAM. IEEE Comput. **22**, 51–64 (1989)
6. Dimitriou, T.: A secure and efficient RFID protocol that could make big brother (partially) obsolete. In: Proceedings of IEEE PERCOM (2006)
7. EPC Radio-Frequency Identity Protocols Class-1 Gen-2 UHF RFID Protocol for Communications at 860MHz-960MHz, EPCglobal (2011). Available at http://www.epcglobalinc.org/uhfclg2

8. High Memory On Metal UHF RFID Tags: Available at http://www.oatsystems.com/OAT_Xerafy_RFID_Aerospace_2013/media/High-Memory-Tag-Guide.pdf

9. Juels, A., Weis, S.A.: Defining strong privacy for RFID. In: IEEE PerCom Workshops, pp. 342–347 (2007)

10. Li, T., Luo, W., Mo, Z., Chen, S.: Privacy-preserving RFID authentication based on cryptographical encoding. In: Proceedings of IEEE INFOCOM (2012)

11. Lu, L., Han, J., Hu, L., Liu, Y., Ni, L.: Dynamic key-updating: privacy-preserving authentication for RFID systems. In: Proceedings of IEEE PERCOM (2007)

12. Lu, L., Han, J., Xiao, R., Liu, Y.: ACTION: breaking the privacy barrier for RFID systems. In: Proceedings of IEEE INFOCOM (2009)

13. Lu, L., Liu, Y., Li, X.: Refresh: weak privacy model for RFID systems. In: Proceedings of IEEE INFOCOM (2010)

14. Ohkubo, M., Suzuki, K., Kinoshita, S.: Efficient hash-chain based RFID privacy protection scheme. In: ICUCU, Workshop Privacy (2004)

15. Pais, S., Symonds, J.: Data storage on a RFID tag for a distributed system. Int. J. UbiComp. **2**(2), 26–39 (2011)

16. Ranasinghe, D.C., Cole, P.H.: An evaluation framework. In: Networked RFID Systems and Lightweight Cryptography, Chap. 8. Springer, Berlin (2008)

17. Rukhin, A., Soto, J., Nechvatal, J., Smid, M., Barker, E., Leigh, S., Levenson, M., Vangel, M., Banks, D., Heckert, A., Dray, J., Vo, S. III, L.E.B.: A Statistical Test Suite for Random and Pseudorandom Number Generators for Cryptographic Applications. National Institute of Standards and Technology, Gaithersburg, MD (2010)

18. Singular Value Decomposition: Available at http://en.wikipedia.org/wiki/Singular_value_decomposition

19. Tsudik, G.: Ya-trap: Yet Another Trivial RFID Authentication Protocol. In: Proceedings of IEEE PerCom (2006)

20. Weis, S., Sarma, S., Rivest, R., Engels, D.: Security and Privacy Aspects of Low-cost Radio Frequency Identification Systems. Lecture Notes in Computer Science. Springer, New York (2004)

Chapter 4
Identifying State-Free Networked Tags

Traditional RFID technologies allow tags to communicate with a reader but not among themselves. By enabling peer communications between nearby tags, the emerging networked tags represent a significant enhancement to today's RFID tags. They support applications in previously infeasible scenarios where the readers cannot cover all tags due to cost or physical limitations. This chapter introduces a fundamental problem of identifying networked tags. To prolong the lifetime of networked tags and make identification protocols scalable to large systems, energy efficiency and time efficiency are most critical. We reveal that the traditional contention-based protocol design will incur too much energy overhead in multihop tag systems, while a reader-coordinated design that significantly serializes tag transmissions performs much better. In addition, we show that load balancing is important in reducing the worst-case energy cost to the tags, and we present a solution based on serial numbers.

The rest of this chapter is organized as follows. Section 4.1 presents the system model and the problem statement. Section 4.2 discusses the related work. Section 4.3 describes the contention-based ID collection protocol. Section 4.4 introduces the serialized ID collection protocol. Section 4.5 presents two techniques to improve the time efficiency of serialized ID collection. Section 4.6 evaluates the performance of our protocols by simulations. Section 4.7 gives the summary.

4.1 System Model and Problem Statement

4.1.1 Networked Tag System

A networked tag system consists a reader and a large number of objects, each of which is attached with a tag. We will use tag, node, and networked tag

© Springer International Publishing AG 2016

M. Chen, S. Chen, *RFID Technologies for Internet of Things*,

Wireless Networks, DOI 10.1007/978-3-319-47355-0_4

interchangeably in the sequel. Each tag has a unique ID that identifies the object it is attached to. The reader also has a unique ID that differentiates itself from the tags.

A networked tag system is different from a traditional RFID system with a fundamental change: Tags near each other can directly communicate. This capability allows a multihop network to be formed amongst the tags. Developed at Columbia University recently [2], networked tag prototypes can communicate using variants of CSMA and slotted ALOHA. The transmission range of inter-tag communications is usually short, about 1–10 m [5]. But the reader is a more powerful device, and its transmission range can be much larger. Tags that can perform direct two-way communicate with a node form the neighborhood of the node.

Networked tags are expected to carry sufficient internal energy for long-term operations or have the capability of harvesting energy from the environment where they are deployed. Tags of the highest energy demand are located in the reader's neighborhood (i.e., coverage area) because they have to relay the information from all other tags as the data converge towards the reader. Fortunately, these tags can be powered by the reader's radio waves, similar to what today's passive RFID tags do; their energy supply is ensured. In contrast, tags that are beyond the reader's coverage need to use their own energy. The operations of these tags must be made energy-efficient.

The reader and the tags in the system form a connected network. In other words, there exists at least one path between the reader and any tag such that they can communicate by transmitting data along that path. Tags that are not reachable from the reader are not considered to be in the system.

4.1.2 Problem Statement

The problem of tag identification is for a reader to collect IDs from all networked tags that can be reached by the reader over multiple hops with the help of intermediate tags relaying the IDs of tags that are not in the immediate coverage area of the reader. Our goal is to develop tag identification protocols that are efficient in terms of energy cost and protocol execution time. We will consider both average energy cost per tag and maximum energy cost among all tags in the system. The average energy cost is an overall measurement of energy drain across the whole system, and the maximum energy cost is a measurement for the worst hot spot which may cause power-exhausted tags and network partition.

4.1.3 State-Free Networked Tags

There are two types of networked tags. The stateful networked tags maintain network state such as neighbors and routing tables and update the information to keep it up-to-date. These tags resemble the nodes in a typical sensor network.

On the contrary, for the purpose of energy conservation, the state-free tags do not maintain any network state prior to operation, which makes them different from traditional networks, including sensor networks—virtually all literature on data-collecting sensor networks assume the stateful model, where the sensor nodes maintain information about who are their neighbors and/or how to route data in the network. We consider state-free networked tags, not only because there is little prior work on this type of networked nodes, but also because it makes more sense for the tag identification problem: First, establishing neighborship and then routing tables across the network is expensive and may incur much more overhead than tag identification itself, which only requires each tag to deliver one number (its ID) to the reader. Second, maintaining the neighbor relationship and updating the routing tables (as tags may move between operations) require frequent network-wide communications, which is not worthwhile for infrequent operation of tag identification.

It is challenging to design an identification protocol for state-free networked tags. First, because power is a scarce resource for tags, the protocol must be energy-efficient in order to reduce the risk of network failure caused by energy depletion. Second, we should also make the protocol time-efficient so that it can scale to a large tag system where the communication channel works at a very low rate for energy conservation. Third, in order to eliminate overhead of state maintenance and thus conserve energy, tags are assumed to be state-free, which means that they do not know who are their neighbors and there is no existing routing structure for them to send IDs to the reader.

4.1.4   System Model

What makes tags attractive is their simplicity. There is no specification on how simple future networked tags should be, but it is safe to say that we will always prefer protocol designs that achieve comparable efficiency with less hardware requirement. Generally speaking, each tag has very limited energy, memory, and computing resources. In this chapter, we do not require tags to implement GPS, any localization mechanism, or other complex functions. We consider state-free tags, which do not spend energy in maintaining any state information prior to operation.

Since each tag is only equipped with a single transceiver, it cannot perform transmission and reception simultaneously. Assume that the reader and tags cannot resolve collided signals. Therefore, a node can successfully receive the transmission only if there is only one neighbor transmitting.

For state-free tags, there is no mechanism (such as frequent beacon exchange between neighbors) that keeps track of the changes in network topology in real time. We assume that the tags are stationary during the operation of tag identification. For example, in a warehouse, the daily tag identification may be performed automatically in after-work hours when objects are not moved around. During the daytime between the previous identification and the next one, objects can still be

Table 4.1 Notations

Symbols	Descriptions
$T/T'/T_i$	Networked tag in the system
N	Number of networked tags in the system
f	Frame size used by the reader
s/s'	Serial number
R	Transmission range of the reader
r	Inter-tag transmission range
D_i	Average children degree of tier-i tags
L_i	Average load factor of tier-i tags
N_i	Number of tags on the ith tier
ρ	Distribution density of networked tags
λ	Time frame size
$RQST_1$	Request to instruct tags to report their IDs
$RQST_2$	Request to instruct a designated tag to collect IDs

freely moved around. In case that the identification operation needs to be performed during the daytime, we need to design a protocol that takes as little time as possible to avoid significant interruption to other warehouse operations due to the stationary requirement at the time of identification.

To conserve energy, networked tags are likely configured to sleep and wait up periodically for operations. After wake-up, a tag will listen for a request broadcast from the reader into the network, which either puts the tag back to sleep or asks the tag to participate in an operation such as reporting its ID. The broadcast request will serve the purpose of loosely re-synchronizing the tag clock. The reader will time its next request a little later than the timeout period set by the tags to compensate for the clock drift and the clock difference at the tags due to broadcast delay. The exact sleep time of the tags and the inter-request interval of the reader should be set empirically based on application needs and physical parameters of the tags.

Notations used in this chapter are given in Table 4.1 for quick reference.

4.2   Related Work

The tag identification protocols for traditional RFID systems can be broadly classified into two categories: ALOHA-based [12, 15] and tree-based [9, 14]. To run an ALOHA-based identification protocol, the reader first broadcasts a query, which is followed by a slotted time frame. Each tag randomly picks a time slot in the frame to report its ID. Collision happens if a slot is chosen by multiple tags. Tags not receiving positive acknowledgements from the reader will continue participating in the subsequent frames. The dynamic frame slotted ALOHA (DFSA) [10, 11] adjusts the frame size round by round.

The tree-based protocols organize all IDs into a tree of ID prefixes. Each in-tree node has two child nodes that have one additional bit, "0" or "1". The tag IDs are leaves of the tree. The reader walks through the tree. As it reaches an in-tree node, it queries for tags with the prefix represented by the node. When multiple tags match the prefix, they will all respond and cause collision. Then the reader moves to a child node by extending the prefix with one more bit. If zero or one tag responds (in the one-tag case, the reader receives an ID), it moves up in the tree and follows the next branch.

To further improve the identification efficiency, network coding and interference cancelation techniques are used to help the reader recover IDs from collided signals [7, 16].

4.3   Contention-Based ID Collection Protocol for Networked Tag Systems

We are not aware of any existing data collection protocol specifically designed for the state-free model which makes sense in the domain of tags but was not adopted in the mainstream literature of sensor networks or other types of wireless systems. However, it is not hard to design an ID collection protocol for networked tags based on techniques known in existing wireless systems. For example, in this section, we will follow an obvious design path based on broadcast, spanning tree, and contention-based transmission. The resulting protocol will be used as a benchmark for performance comparison (since there is no prior work on identifying networked tags). In the next section we will point out that the obvious techniques are, however, inefficient and other less-obvious design choices can produce much better performance.

4.3.1   Motivation

One straightforward approach for tags to deliver their IDs to the reader is through flooding: As each tag broadcasts its ID and every other ID it receives for the first time into its neighborhood, the IDs will eventually reach the reader. However, flooding causes a lot of communication overhead. In addition, each tag has to store the IDs that it has received in order to avoid duplicate broadcast. Due to the nature of flooding, it means that eventually each tag will store all IDs in the system, which demands too much memory.

Another approach is to ask tags to discover their neighbors and run a routing protocol to form routing paths towards the reader right before sending the IDs (even though the tags are state-free prior to operation). However, as the number of neighbors can be in hundreds in a packed system, the overhead of doing so will be high, considering that only one ID per tag will be delivered.

As the above two approaches do not work well, our idea is to establish routing paths for free. For a reader to begin the tag identification process, it needs to broadcast a request to all tags. We can make extra use of this network-wide broadcast to piggyback the function of establishing a spanning tree that covers all tags, with the reader at the root of the tree. This tree will be used for transmitting the IDs to the reader. We use the ALOHA protocol to resolve the contention among concurrent transmissions made by close-by tags.

4.3.2   Request Broadcast Protocol

The classical broadcast protocol is for each node to transmit a message when it receives the message for the first time. But it becomes more complicated to guarantee that all nodes receive the message: If each node knows its neighbors, it may keep transmitting the message until receiving acknowledgements from all neighbors. However, more care must be taken if the nodes do not know their neighbors. Below we briefly describe a request broadcast protocol (RBP) that guarantees delivering a request from the reader to all state-free tags.

To initiate tag identification, the reader broadcasts a request notifying the tags to report their IDs. The request initially carries the reader's ID, which will later be replaced with a tag's ID when the tag forwards the request to others. The state transition diagram of the protocol is depicted in Fig. 4.1, which is explained below.

State of Waiting for Request   Each tag begins in this state and takes action based on one of three possible events.

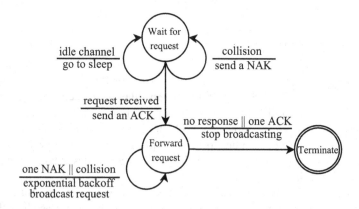

Fig. 4.1 State transition diagram of the RBP protocol. Each *circle* is a state, and each *arrow* is a transition, where the event triggering the transition is *above the line* and the action is *below the line*

1. Idle Channel: The channel is idle, i.e., no neighbor is transmitting anything.
2. Request Received: Only one neighbor is forwarding the request, so the tag can receive the request correctly.
3. Collision: Multiple neighbors are forwarding the request, resulting in a collision.

In event (1), the tag does nothing. In event (2), the tag will acknowledge the sender that it has successfully received the request with an ACK. Meanwhile, it extracts the ID from the request and saves it as its parent. After that, it moves to the state of Forwarding Request. As we will see shortly, it is not important whether the ACK is correctly received by the sender of the request or not. The sender will know that all its neighbors have received the request when it does not hear any response (since the neighbors all move to the state of Forwarding Request). In event (3), the tag cannot resolve the collided. It sends a negative acknowledge (NAK) and stays in the state of Waiting for Request.

State of Forwarding Request To ensure that the request will be propagated across the network, each tag having received the request will keep broadcasting it with exponential backoff upon collision until all its neighbors receive the request. Each time after the tag broadcasts the request (which carries the tag's ID), there are three possible events:

1. No Response: No response is received from any neighbor.
2. One ACK/NAK: Only one ACK/NAK response is received.
3. Collision: Multiple ACK/NAK responses are sent by the neighbors, leading to a collision.

Recall that any neighbor in the state of Wait for Request will respond either ACK or NAK regardless of whether it can successfully receive the request or not. Event (1) must mean that all the neighbors have already received the request and moved to other states. In this case, the tag does not need to broadcast the request any more. If no response is heard after broadcasting the request for the first time, the tag knows it has no child and it is therefore a leaf node in the spanning tree. In event (2), if a single ACK is received, the tag knows that all its neighbors now have received the request. Hence, it can stop broadcasting the request. If a single NAK is received, the tag knows that there must have been collision at a neighbor, which did not receive the request successfully. Therefore, the tag should perform an exponential backoff to avoid continuous collision in the channel. In event (3), the tag cannot resolve the received ACK/NAK correctly and it also performs an exponential backoff. As an example, Fig. 4.2 illustrates the spanning tree built in a networked tag system after it executes RBP, where the reader has an ID 0.

The wireless transmissions in RBP can be implemented either based on unslotted ALOHA or based on slotted ALOHA. Slotted ALOHA is more efficient but requires the tags to synchronize their slots. When the reader transmits its request to nodes in its neighborhood, the preamble of the transmission provides the clock and slot synchronization. Similarly, when a distant tag receives the request for the first time from another tag, the preamble of the latter synchronizes the clock and slot.

Fig. 4.2 An example of a spanning tree built by RBP, where the ID of tag T_i is i. Each *dotted circle on the left* gives the neighbors of a tag at the *center of the circle*

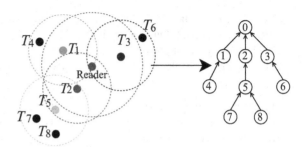

Theorem 4. *Every tag will receive a copy of the request sent out by the reader under RBP.*

Proof. To prove by contradiction, let's assume there exists a tag T that does not receive the request after executing RBP. It must be true that none of its neighbors has received the request. Otherwise, according to the protocol, any neighbor having received the request would continue broadcasting the request until T receives it and acknowledges its receipt—each time the request is transmitted, if T does not receive the request successfully, it will respond NACK, causing the sender to retransmit. By the same token, the neighbors of any T's neighbor must not receive the request. Applying this argument recursively, all nodes reachable from T must not receive the request. By the assumption that the network is connected, at least one neighbor T' of the reader is reachable from T. Therefore, T' must not receive the request. This contradicts to the fact that T' is located in the reader's coverage area and should receive the request at the very beginning when the reader broadcasts the request for the first time. Therefore, the theorem must hold.

4.3.3   ID Collection Protocol

When a tag transmits its ID, it will include its parent's ID in the message, such that the parent node will receive it while other neighbors will discard the message. This unicast transmission is performed based on the classical ALOHA with acknowledgement and exponential backoff to resolve collision. The parent node will forward the received ID to its parent, and so on, until the ID reaches the reader.

The execution of ICP is performed in parallel with RBP: Once a tag knows its parent ID from RBP, it will begin transmitting its ID to the parent. When a tag needs to forward both an ID for ICP and a request for RBP, we give priority to ID forwarding because it is easier for unicast to complete.

Theorem 5. *The reader will receive the IDs of all tags in the system after the execution of ICP.*

Proof. From Theorem 1, each tag is guaranteed to receive the request and therefore find a parent (from which the request is received). Consider an arbitrary tag T. According to the design of ICP, the ID of T will be sent to its parent until positively acknowledged. The parent will forward the ID to its parent, and as this process repeats, the ID will eventually reach the reader at the root of the spanning tree.

4.4 Serialized ID Collection Protocol

4.4.1 Motivation

The contention-based protocol ICP allows parallel transmissions by non-interfering tags through spatial channel reuse. In the conventional wisdom, this is an advantage. However, we find in our simulations that the contention-based protocol performs poorly for tag identification. The reason is that although parallel transmissions are enabled among the tags in the network, the reader can only take one ID at a time. Essentially, the operation of ID collection is serialized at the reader, regardless of how much parallelism is achieved inside the network of tags. Furthermore, the parallelism is actually harmful because the more the IDs are crowded to the reader in parallel, the more the contention is caused at the reader, resulting in many failed transmissions due to collision, which translates into high energy cost and long protocol execution time. When tags are densely deployed, this problem can severely degrade the system performance. With this observation, we take a different design path by trying to partially serialize the tag transmissions, such that only a (small) portion of tags will attempt to transmit at any time. By lessening the level of contention, we see a drastic performance improvement. Another serious problem of RBP/ICP is that the spanning tree is unbalanced, causing significantly higher energy expenditure by some tags than others. This problem of biased energy consumption and a solution will be explained in details later.

4.4.2 Overview

We give an overview of our serialized protocol, SICP. The reader begins by collecting IDs in its neighborhood using framed ALOHA. An illustrative example is shown in Fig. 4.3, where the reader collects the IDs from neighbors T_1 through T_4 (which form tier 1), while all other nodes stay idle. When the reader receives a tag's ID (say, T_2) free of collision, T_2 must be the only tag that is transmitting in the whole network. It also means that other neighbors of T_2 can hear the transmission free of collision. These tier-2 nodes, T_5 through T_7, set T_2 as their parent.

After collecting all tier-1 IDs, the reader sequentially informs each tier-1 node to further collect IDs from its children. For example, when the reader informs T_2 to do

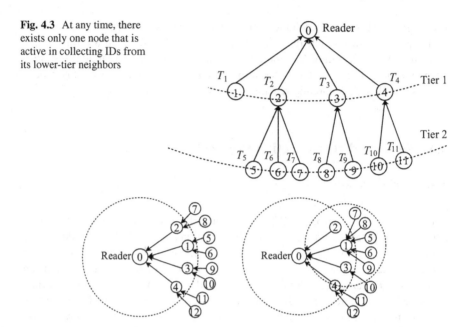

Fig. 4.3 At any time, there exists only one node that is active in collecting IDs from its lower-tier neighbors

Fig. 4.4 *Left plot*: a roughly balanced spanning tree; *Right plot*: a biased spanning tree, where tag 1 delivers its ID to the reader first and a large number of nodes in its neighborhood chooses it as their parent, causing a biased tree. The serial number of a tag is shown inside the *circle*. Each *arrow* represents a child–parent relationship

so, all other tier-1 nodes will stay silent. As T_2 sends out a request for IDs, only its children (T_5, T_6 and T_7) will respond. The same process as described in the previous paragraph will repeat; only this time T_2 takes the role of the reader.

After T_2 collects the IDs of all its children, it will forward the IDs to the reader, which will then move to the next tier-1 node. Once it exhausts all tier-1 nodes, it will move to tier-2 nodes, one by one and tier by tier, until the IDs of all nodes in the network are collected.

Below we will first introduce the problem of biased energy consumption, give a solution, and then describe recursive serialization.

4.4.3 Biased Energy Consumption

When a tag is transmitting its ID to the reader, its neighbors outside of the reader's coverage can overhear the ID. They may use this tag as their parent. As illustrated in the left plot of Fig. 4.4, we prefer a roughly balanced spanning tree where each node serves as the parent for a similar number of children. In reality, however, a tag that delivers its ID to the reader early on will tend to have many more children. An example is given in the right plot of Fig. 4.4. Suppose tag 1 transmits its ID to the

Fig. 4.5 A tag may choose
its parent from multiple
candidates, where *arrows*
represent ID transmissions (or
broadcast)

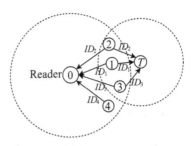

reader first. Overhearing its ID, tags 5–9 will pick tag 1 as their parent. When tag 2 transmits its ID at a later time, no tag will be left to choose tag 2 as parent even though tags 7–8 are in the range of tag 2—recall that they have already chosen tag 1. In this case, tag 1 will have to forward more IDs, resulting in quicker energy drain than others. The severity of the problem grows rapidly with an increasing number of tiers because the numerous children of tag 1 tend to acquire even more numerous children of their own and those IDs will pass through tag 1 to the reader.

Uneven energy consumption causes some tags to run out of energy earlier, which can result in network partition. The same problem also exists for RBP/ICP where tags that receive and forward the request early on during the network-wide broadcast may end up with a large number of children.

We observe that a tag may overhear multiple ID transmissions over time and thus have multiple candidates to choose its parent from, as shown by Fig. 4.5 where T may choose its parent from three tier-1 nodes. Ideally, the tag should choose its parent uniformly at random from the candidates. However, because of collision, each candidate may have to retransmit its ID for a different number of times before the reader successfully receives it. To avoid giving more chance to a candidate that retransmits its ID more times, the tag may keep the IDs of all known candidates to filter out duplicate overhearing. However, a serious drawback of this approach is that the memory cost can be high if a tag has numerous candidates for its parent in a system where tagged objects are packed tightly together. We want to point out that typical tags have very limited memory.

4.4.4 Serial Numbers

We present a solution to biased energy consumption based on serial numbers. In our protocol, each tag will be dynamically assigned a serial number from 1 to N, where N is the number of tags. The reader's serial number is 0.

Let's first consider the reader's neighborhood only; other tiers will be explained later. The reader initiates the protocol by broadcasting an ID collection request denoted by $RQST_1$, carrying its serial number and a frame size f. The request is followed by a time frame of f slots. Each tag that receives the request will set the serial number 0 (i.e., the reader) as its parent and then randomly chooses a slot in the time frame. It waits until the chosen slot to report its ID to the reader. If only one tag selects a certain slot, its ID will be correctly received by the reader, which replies an ACK to the tag in the same slot. The ACK carries the number of IDs that the reader has successfully received so far. This number is assigned as the serial number of the tag; the number is system-wide unique due to its monotonically increasing nature. A tag can be identified either by its ID or its assigned serial number. After receiving the ACK, we require the tag to broadcast the assigned serial number in its neighborhood. Hence, each time slot contains an ID transmission, an ACK transmission, and a serial-number transmission. If the ID transmission is collision-free, so do the other two transmissions. Even though a tag may need to retransmit its ID multiple times due to collision, it will transmit its assigned serial number once, only at the time when an ACK is received.

If the reader observes any collision in the time frame, it will broadcast another request with another time frame to collect more IDs. If no collision is observed, the reader has collected all IDs from its neighborhood and it will perform recursive serialization (to be discussed) to collect IDs outside of its neighborhood.

4.4.5   Parent Selection

Consider an arbitrary neighbor of T, denoted as T', which has not set its parent yet. As illustrated in Fig. 4.6, T' must not be in the reader's neighborhood because the tags in that neighborhood set the serial number 0 as their parent when they receive the request from the reader for the first time. When T' receives a serial number for the first time, it will set the number as its parent, which is subject to change when T' receives more serial numbers from other tags (candidates for parent). Recall that each tag broadcasts its serial number only once. This property allows us to design the following parent selection algorithm (PSA) which guarantees every candidate has an equal chance to be selected as the parent: Each tag maintains two values, its parent and a counter c for the number of candidates having been discovered so far. The counter is initialized to zero. Each time when T' receives a serial number s' from a neighbor, it increases c by one and then replaces the current parent with s' by a probability $\frac{1}{c}$. Using this PSA, we have the following theorem:

Fig. 4.6 T' sets T as its parent

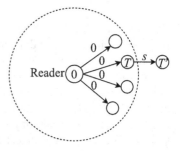

Theorem 6. *Suppose a tag has m candidates for parent. Each candidate has an equal probability of $\frac{1}{m}$ to be chosen as the tag's parent in the end.*

Proof. For the jth $(1 \leq j \leq m)$ discovered candidate, it becomes the final parent only if it replaces the previously selected parent, and is never substituted by the subsequently discovered candidates. Therefore, the probability that it is chosen by the tag as the parent in the end is

$$\frac{1}{j} \prod_{l=j+1}^{m} (1 - \frac{1}{l}) = \frac{1}{m}, \tag{4.1}$$

implying every candidate is equally likely to be the parent.

Another advantage of using the serial number instead of ID for parent identification is that an ID—typically 96 or more bits for RFID tags—is much longer than the size of the serial number, $\lceil \log_2 N \rceil$, where N is the maximum number of tags in a system. For example, even if $N = 1{,}000{,}000$, the serial number is just 20 bits long.

4.4.6   Serialization at Tier Two

After the reader collects all IDs from its neighborhood, each tag in the neighborhood will obtain a unique serial number. Recall that these tags constitute the first tier of the network. The reader then serializes the subsequent ID collection process by sending the serial numbers of tier-1 tags one by one, in order to command the corresponding tag to collect IDs from its neighbors, with other tier-1 tags staying idle.

The reader begins by transmitting another type of request, denoted by $RQST_2$, which includes the serial number 1 and the number s of IDs it has received so far. In response, the tag with the serial number 1, denoted as T_1, transmits an ID collection request $RQST_1$, carrying its own serial number 1 and a frame size f. The request causes the neighbors that are not tier-1 to finalize their parent selection; these nodes are tier-2. Note that some of them may have selected nodes other than T_1 as their parents. Hence, when a tier-2 node receives the request from T_1, only if its chosen parent matches the serial number in the request, it will transmit its ID in the subsequent time frame; otherwise, it can sleep for a duration of f slots. If T_1 correctly receives an ID in a slot from a child T_1', it increases the value of s by one and sends back an ACK with s as the serial number assigned to T_1', which in turn broadcasts its serial number and tier number (i.e., 2) in its neighborhood such that the neighbors at the next tier can discover it as one of their candidates for parent. When a tag sets (or later replaces) its parent, it also sets its tier number as the tier number of its parent plus one; it should never replace its current parent with one whose tier number is larger.

It may take T_1 multiple requests to finish reading all IDs from its children. It then forwards the IDs to the reader. After acknowledging T_1, the reader sends a command to trigger the ID collection process at the next tier-1 tag.

After the reader finishes this process with all tier-1 tags, it has collected the IDs of all tier-2 tags. The reader also has the information to construct a spanning tree covering tier-1 and tier-2 nodes, as illustrated in Fig. 4.3 where the assigned serial numbers are shown inside the circles.

4.4.7   Recursive Serialization

After the reader commands all tier-1 tags one by one to collect the IDs of tier-2 tags, it repeats this serialization process recursively to collect other IDs tier by tier. Suppose the reader has collected the IDs from all tags at tier 1 through tier i and the range of serial numbers at tier i is from x to y. The reader will send an $RQST_2$ to each tier-i tag in sequence. The command includes a concatenation of the serial numbers along the path in the spanning tree from the root (excluded) to that tag, in addition to the number s of IDs that the reader has received so far. For example, for tag 7 in Fig. 4.3, the command will carry two serial numbers, 2 and 7. (Note that since each serial number is of fixed size, there is no ambiguity on interpreting the sequence of serial numbers.)

When the reader broadcasts the command in its neighborhood, any tag receiving the command will extract and compare the first serial number with its own. If the two serial numbers do not match, it discards the command. Otherwise, it further checks whether there are more serial numbers in the command. If so, it broadcasts the remaining command. This process repeats until a tag matches the last serial number in the command. That tag will performs ID collection in a similar way as described in Sect. 4.4.6. The collected IDs will be sent through the parent chain to the reader.

Theorem 7. *The reader will receive the IDs of all tags in the system after the execution of SICP.*

Proof. Proving by contradiction, we assume at least one tag T fails in delivering its ID to the reader. T must not have a parent; we again prove this by contradiction: Assume that T has a parent T'. According to the protocol, for T' to be chosen as a parent, it must either the reader or a node that has already successfully delivered its ID and subsequently broadcast its assigned serial number. Hence, it will receive a command from the reader to collect IDs from its children. After the reader sends a command to T', T' will broadcast requests, free of collision due to serialization, to children until all IDs are collected—which happens when no collision is detected in the time frame after a request. When T' receives the ID of T, if it is not the reader, it will forward the ID to the reader along the path with which its own ID has been successfully delivered, free of collision due to serialization. This contradicts to the assumption that T fails in delivering its ID to the reader. Hence, T does not have a parent.

If T does not have a parent, all of its neighbors must fail in delivering their IDs to the reader because otherwise any successful neighbor would broadcast its serial number according to the protocol, which would result in T having a parent after T receives the serial number.

If all neighbors of T fail in delivering their IDs to the reader, by the same reasoning as above, all their neighbors must fail too. Recursively applying this argument, all tags in the network must fail in delivering their IDs to the reader because the network is connected, which contradicts at least to the fact that the reader's immediate neighbors are able to send their IDs to the reader through the slotted ALOHA protocol that SICP employs. Hence, the theorem is proved.

The pseudo code of SICP for an arbitrary tag T is given in Protocol 1.

Protocol 1 Serialized ID collection at tag T

 1: **if** T has not reported its ID **then**
 2:     **if** T receives a serial number from a neighbor **then**
 3:         **if** T has not determined its parent **then**
 4:             T executes PSA for parent selection;
 5:         **end if**
 6:     **else if** T receives a $RQST_1$ request **then**
 7:         **if** T has not determined its parent **then**
 8:             T sets the candidate parent to its parent;
 9:         **end if**
10:         **if** the request is sent from T's parent **then**
11:             T randomly picks a slot in the following time frame to report its ID;
12:         **else**
13:             T sleeps during the following time frame;
14:         **end if**
15:     **end if**
16: **else if** T receives a $RQST_2$ request **then**
17:     **if** the designated tag in the request is T's descendant **then**
18:         T forwards the request to the target child;
19:     **else if** the designated tag in the request is T itself **then**
20:         T performs ID collection from its children;
21:         T forwards the collected IDs to its parent;
22:     **else**
23:         T discards the request;
24:     **end if**
25: **else if** T receives a forwarding request **then**
26:     **if** the request is sent from its child **then**
27:         T receives and forwards the IDs to its parent;
28:     **else**
29:         T sleeps when its neighbor is forwarding IDs;
30:     **end if**
31: **end if**

4.4.8   Frame Size

When the reader or a tag tries to collect the IDs in its neighborhood, its request carries a frame size f. Let n be the number of tags that are children of the reader or tag sending the request. It is well known that the optimal frame size should be set as n, such that the probability of each slot carrying a single ID (without collision) can be maximized. This can be easily seen as follows: Consider an arbitrary slot. The probability p that one and only one tag chooses this slot to transmit is

$$p = \binom{n}{1}\frac{1}{f}\left(1-\frac{1}{f}\right)^{n-1} \approx \frac{n}{f}e^{-\frac{n-1}{f}} \approx \frac{n}{f}e^{-\frac{n}{f}} \tag{4.2}$$

when n is large. To find the value of f that maximizes p, we take the first-order derivative of the right side and set it to zero. Solving the resulting equation, we have

$$f = n, \tag{4.3}$$

which means the maximal value of p is e^{-1}. In subsequent requests, as more and more IDs have been collected, fewer and fewer tags are transmitting their IDs and the frame sizes should be reduced accordingly.

However, we do not know n. There are numerous estimation methods for n [3, 6, 13], which are, however, intended for a system with a large number of tags, in tens of thousands. It is known that these estimation methods will actually be inefficient if they are applied to a relatively small number of tags such as a couple of thousands or fewer [8]; if the number of tags is very small, the estimation time can be much larger than the time it takes to complete the tag identification task itself. In the context of this chapter, we expect the number of children of the reader or any tag is relatively small. Hence, it is not worthwhile to add the overhead of a separate component for estimating n before the reader (tag) begins collecting IDs from its neighborhood.

Our solution is to estimate the value of n iteratively from the frame itself without incurring additional overhead. Initially, we set f to be a small constant λ in the first request. We double the value of f in each subsequent request until there exists at least one empty slot that no tag chooses. From then on, we will estimate the number of n and set the frame size accordingly in the subsequent requests. Without the loss of generality, suppose we want to determine the frame size for the ith request. Let f_j be the frame size used in the jth request, $1 \leq j < i$. After the jth request, let c_j, s_j, and e_j be the numbers of slots that are chosen by multiple tags (collision), a single tag, and zero tag, respectively. Let m_j be the number of IDs that are successively collected after the jth request. All these values are known to the reader (tag). The process for a tag to randomly choose a slot in a time frame can be cast into bins and balls problem [4]. In the jth frame, $n - m_{j-1}$ tags (balls) are mapped to f_j slots (bins). The total number of different ways for putting $n - m_{j-1}$ balls to f_j bins is $f_j^{n-m_{j-1}}$. The number of ways for choosing e_j bins from f_j bins and let them be empty is $\binom{f_j}{e_j}$.

In addition, the number of ways for choosing s_j balls from $n - m_{j-1}$ balls and putting each of them into one of the remaining $f_j - e_j$ bins is $\binom{f_j - e_j}{s_j}\binom{n-m_{j-1}}{s_j}(s_j!)$. Finally, the remaining $n - m_{j-1} - s_j$ balls should be thrown into the remaining c_j bins, each containing at least two balls (collision slots). We first choose $2c_j$ balls and put two balls into each of the c_j bins, which includes $\binom{n-m_{j-1}-s_j}{2c_j}\frac{(2c_j)!}{2^{c_j}}$ possibilities. After that, the remaining $(n - m_{j-1} - s_j - 2c_j)$ balls can be put into any of the c_j bins, which involves $(n - m_{j-1} - s_j - 2c_j)^{c_j}$ different ways. Therefore, the likelihood function for observing these values is

$$L(n) = \prod_{j=1}^{i-1} \frac{\binom{f_j}{e_j}\binom{f_j - e_j}{s_j}\binom{n-m_{j-1}}{s_j}(s_j!)\binom{n-m_{j-1}-s_j}{2c_j}\frac{(2c_j)!}{2^{c_j}}}{f_j^{n-m_{j-1}}} \times (n - m_{j-1} - s_j - 2c_j)^{c_j}.$$

(4.4)

The estimate of n is the value that maximizes L. Let this value be $\hat{n}$, which can be found through exhaustive search since the range for n is limited in practice, rarely going beyond tens of thousands. For the ith request, we set the frame size to be $\hat{n} - m_{i-1}$.

The above estimator follows the general principle originally seen in [6], but it takes the information of c_j, s_j, and e_j all in the same estimator, whereas the estimators in [6] use either c_j or e_j.

As our analysis will show, except for the reader, the average number of children per tag is typically very small (less than 2) for a randomly distributed tag network. In this case, if we set the initial frame size λ to 4, the chance is high that a tag successfully collect all IDs from it children in the first time frame. Therefore, only the reader needs to use (4.4) to estimate the number of its children, while the tags can just set the frame size to a small constant to avoid the computation overhead.

4.4.9 Load Factor Per Tag

We analyze the work load of each tag in terms of how many children and descendants it has to handle. While our load balancing approach is designed for any tag distribution, to make the analysis tractable, we assume here that tags are evenly distributed in an area with density ρ, and the tags whose distances from the reader are no larger than R form the first tier, while those whose distances from the reader are greater than $R + (i-2)r$ but smaller than $R + (i-1)r$ form the ith ($i \geq 2$) tier of the network, where the transmission ranges of the reader and a tag are R and r, respectively, with $R \geq r$. For example, Fig. 4.7 presents a network with three tiers. The number N_i of tags in the ith tier is estimated as

$$N_i = \rho \times (\pi \times (R + (i-1)r)^2 - \pi \times (R + (i-2)r)^2)$$
$$= \pi\rho(2Rr + (2i-1)r^2).$$

(4.5)

Fig. 4.7 An illustration of a
network with three tiers of
tags

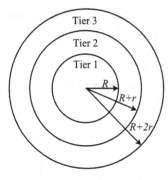

Table 4.2 The values of D_i
with $R = 3r$

i	1	2	3	4	5	6	7	8	9
D_i	1.3	1.2	1.2	1.2	1.1	1.1	1.1	1.1	1.1

One exception is that N_1 computed from (4.5) actually includes only the portion of
tier-1 tags whose distances from the reader are larger than $R - r$; these are the tags
that can serve as parents for tier-2 tags.

The children degree of tier-i tags, denoted by D_i, is defined as the average number
of children that a tier-i tag has. Because tags at the ith tier only serve as parents for
tags at the $(i + 1)$th tier, we have

$$D_i = \frac{N_{i+1}}{N_i} = \frac{2R + (2i + 1)r}{2R + (2i - 1)r} = 1 + \frac{1}{\frac{R}{r} + (i - \frac{1}{2})}. \tag{4.6}$$

We have $R \gg r$ because the reader can transmit at a much higher power level and
it has much more sensitive antenna. This makes the values of D_i very small. For
example, if $R = 3r$, Table 4.2 shows the values of D_i, $1 \leq i < 10$, which are smaller
than 1.3 and quickly converge towards 1 as i increases. The values in the table will
be even smaller if $R > 3r$.

The load factor of tier-i tags, denoted as L_i, is defined as the average number of
IDs that a tier-i tag has to forward, including the IDs of its tier-$(i + 1)$ children as
well as other IDs that its children collects from their descendants. L_i is equal to the
total number of tags beyond the ith tier divided by the number of tags at the ith tier.

$$L_i = \frac{\sum_{j=i+1}^{l} N_j}{N_i} = \frac{\sum_{j=i+1}^{l} 2R + (2j - 1)r}{2R + (2i - 1)r}$$
$$= \frac{2(l - i) + \frac{r}{R}(l^2 - i^2)}{2 + (2i - 1)\frac{r}{R}}, \tag{4.7}$$

where l is the total number of tiers and $i < l$. When $R = 3r$ and $l = 10$, Table 4.3
shows the values of L_i, $1 \leq i < 10$, which are surprisingly small. Because tier-1 tags
can be powered by the radio wave from the reader, we are only concerned with the

Table 4.3 The values of L_i with $R = 3r$ and $l = 10$

i	1	2	3	4	5	6	7	8	9
L_i	21.9	16.0	12.1	9.2	7.0	5.2	3.6	2.3	1.1

power consumption of tags at other tiers. The tags at tier 2 have to forward more IDs than those at outer tiers. From the table, a tier-2 tag forwards just 16 IDs on average, which is modest overhead, considering that there are eight more tiers beyond tier 2.

While the average is modest, the worst-case load factor is also important when we evaluate overhead. SICP is designed to evenly distribute the work load among tags by balancing the spanning tree, so that tags at a certain tier have similar numbers of children (or descendants), which translate to similar children degrees (or load factors). We will study the worst-case children degree and load factor by simulations.

4.5 Improving Time Efficiency of SICP

The serialized ID collection of SICP eliminates most simultaneous transmissions for the purpose of reducing collision, which in turn may degrade the time efficiency. In this section, we explore potential ways to improve the time efficiency of SICP. The process for tags to execute SICP includes three steps: (1) the reader sends a request to a designated tag to perform ID collection; (2) the designated tag collects IDs from its children; and (3) the collected IDs are forwarded to the reader. In step (2), IDs are collected using the frame slotted ALOHA protocol, and the optimal frame size is given by (4.3). In practice, the frame size used by tags may be set to a small constant since each tag only has a few children, thereby reducing tags' computation overhead for optimizing the frame size. Hence, our objective here is to improve time efficiency of step (1) and step (3).

4.5.1 Request Aggregation

Recall that each tag has to receive an $RQST_2$ request from the reader before collecting IDs from its children. The request is forwarded over multiple hops along the path from the reader to the tag. We first use an example to illustrate the idea of request aggregation. Consider a subtree in Fig. 4.3 that consists of the reader, T_2, T_5, T_6, and T_7. As shown in the upper half of Fig. 4.8, the request to T_5 should be forwarded over two hops reader$\rightarrow T_2$ and $T_2 \rightarrow T_5$. Similarly, the request targeted to T_6 or T_7 needs to be forwarded two times as well. Suppose a one-hop transmission requires one slot. It requires six slots in total to forward all three requests to T_5, T_6, and T_7. We observe that all three requests must be first forwarded to T_2, the common parent of T_5, T_6, and T_7. The reader$\rightarrow T_2$ transmission is carried out three times,

Fig. 4.8 An illustration of using request aggregation to improve time efficiency in SICP

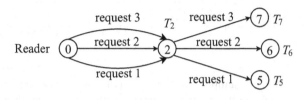

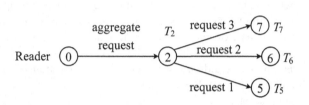

which is redundant and unnecessary. We can indeed aggregate the three requests to a single one to avoid redundant transmissions. As shown in the bottom half of Fig. 4.8, instead of sending separate commands to T_5, T_6, and T_7 individually, the reader first sends an aggregate request to T_2, and T_2 then sends requests to the three children to perform ID collection in sequence. As a result, the total number of slots for forwarding the requests can be reduced to four. In the aggregate request, the reader can include the serial numbers of T_2's children such that T_2 does not need to remember who are its children. Alternatively, we can slightly modify the SICP protocol by asking each tag to record the starting serial number of its children and the number of children its has, which can be easily achieved while performing ID collection. With those two values, each tag can recover all serial numbers of its children when necessary.

Suppose T is a tier-i tag and it has m children. In the original design of SICP, it takes $(i+1)$ slots to forward an $RQST_2$ request to a child of T, which is a tier-$(i+1)$ tag. Hence, the total time cost t_f for forwarding a request to every child of T is

$$t_f = (i+1) \times m. \tag{4.8}$$

After applying the technique of request aggregation, only one aggregate request needs to be sent to T. Therefore, the time cost is reduced to

$$t'_f = i + m. \tag{4.9}$$

4.5.2 ID-Transmission Pipelining

After collecting all IDs from its children, a tag forwards the collected IDs to its parent, which may take multiple slots. Only after receiving all IDs from the tag will the parent start to forward those received IDs. This process continues until the IDs are finally delivered to the reader. Consider a tier-i tag. Suppose it has m children whose IDs are collected, and k IDs can be transmitted in each slot. The time cost t_b, in number of time slots needed to forward the IDs to the reader, is approximately

$$t_b = i \times \frac{m}{k}. \qquad (4.10)$$

This completely serialized way of ID delivery is, however, not time-efficient since only one tag is allowed to transmit at any time. We want to exploit simultaneous transmissions among non-interfering tags through spatial channel reuse. Before introducing the idea of transmission pipelining, we first prove the following theorem:

Theorem 8. *If tag T is an ancestor node but not the direct parent node of tag T' in the spanning tree built by SICP, T' must not be a neighbor of T.*

Proof. Proving by contradiction, we assume that T' is a neighbor of T. Denote the parent of T' as T_p. Recall that a tag will determine its parent when receiving the first ID collect request. Let t' be the time when T' determines its parent, and t be the time when T broadcasts the ID collection request for the first time. Because T' may hear a request for the first time from another node, we must have

$$t' \leq t. \qquad (4.11)$$

Let t_p be the time when T_p successfully delivers its ID. Since T is also an ancestor node of T_p, the delivery of T_p's ID must happen after T sends out its ID connection request. Hence, $t < t_p$. From (4.11), we have

$$t' < t_p. \qquad (4.12)$$

We know that T_p is selected as the parent of T' at time t' (upon the receipt of an ID collection request). In order to select T_p as its parent, T must know the existence of T_p earlier, which happens at time t_p (when T_p successfully delivers its ID to its parent). Namely, it is necessary that

$$t_p < t', \qquad (4.13)$$

which contradicts to (4.12) and thus completes the proof.

Based on Theorem 8, we can conclude that for an arbitrary ID transmission from T'' to T, the transmissions from any of T's ancestor nodes that are not its parent will cause no interference to T's receipt of transmission by T''. We use an example in Fig. 4.9 to illustrate the idea of ID-transmission pipelining. Suppose node 6 has collected the IDs from its children and begins forwarding the IDs to the reader along the path $6 \rightarrow 5 \rightarrow 4 \rightarrow 3 \rightarrow 2 \rightarrow 1 \rightarrow 0$. Instead of forwarding all collected IDs to its parent node 5 at once, which is the method adopted by SICP, node 6 first uses only one slot to transmit some of the IDs (assuming they cannot all fit in one slot) to

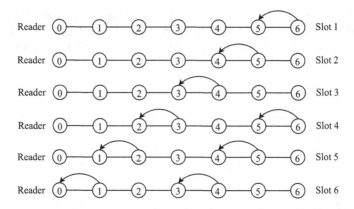

Fig. 4.9 Transmission pipelining of the IDs collected by node 6. Each *arrow* represents one transmission of IDs from a node to its parent in a given slot

its parent, as shown by slot 1 in Fig. 4.9. In the following two slots, only node 5 and node 4 forward the received IDs, respectively, while node 6 does not perform any transmissions to avoid collision. In slot 4, when node 3 forwards the received IDs to node 2, node 6 will transmit another slot of IDs to node 5. According to Theorem 8, node 3 is not a neighbor of node 5, and thus its transmission will not interfere with node 5's receipt of transmission from node 6. In summary, node 6 can perform one transmission every three slots in this example. This is also true for other nodes in the path from node 6 to the reader. The parallel transmissions effectively produce a transmission pipeline in delivering the IDs to the reader.

We generalize the technique of transmission pipelining for any tag in the system as follows:

1. After a tier-1 tag collects IDs from its children, the tag will directly deliver the IDs to the reader in continuous slots without waiting.
2. After a tier-2 tag T collects IDs from its children, the tag performs one ID transmission every two slots, allowing its parent to forward the received IDs to the reader in the time slot immediately following each slot when T transmits.
3. After a tag T at tier 3 or higher collects IDs from its children, the tag performs one ID transmission every three slots to support pipelining. When any tag on the path from T to the reader receives IDs in a slot, it will transmit the received IDs in the next slot.

Therefore, the time cost t'_b for a tier-i tag to forward m collected IDs to the reader with transmission pipelining is approximately

$$t'_b = \begin{cases} i \times \frac{m}{k} & \text{if } 1 \leq i \leq 2 \\ 3 \times (\frac{m}{k} - 1) + i & \text{if } i \geq 3 \end{cases} \tag{4.14}$$

It is easy to prove that $t'_b \leq t_b$. Transmission pipelining can improve the time efficiency of ID collection, particularly for tags with large tier numbers.

We denote the SICP with request aggregation and transmission pipelining as p-SICP in the sequel.

4.6 Evaluation

4.6.1 Simulation Setup

There is no prior work on tag identification for networked tag systems.[1] But known techniques such as broadcast and contention-based transmission widely used in other wireless systems can be used to design a state-free tag identification protocol, CICP, which we will use as a benchmark for comparison. We evaluate the performance of CICP, SICP, and p-SICP to demonstrate three major findings that (1) although the ALOHA-based protocols are very successful in other wireless systems (including RFID systems), they are not suitable for networked tag systems, and that (2) serialization can significantly improve the tag identification performance, and (3) the techniques of request aggregation and ID-transmission pipelining can significantly improve the time efficiency of serialized ID collection.

Three performance metrics are used: (1) execution time measured in number of time slots, (2) average and maximum numbers of bits sent per tag, and (3) average and maximum numbers of bits received per tag. The last two are indirect measures of energy cost, where tier-1 tags are excluded because they can be powered by the reader's radio waves. Computation by tags in the two protocols is very limited. Most energy is spent on communication. The amount of communication data serves as an indirect means to compare different protocols. For example, if tags in one protocol receive and send far more than those in another protocol, it is safe to say that the first protocol costs more energy than the second.

We vary the number N of tags in the system from 1000 to 10,000 at steps of 1000. The tags are randomly distributed in a circular area with a radius of 50 m unless an explicit parameter is specified. The reader, whose communication range R is set to 25 m, is located at the center of the area. For each tag, its inter-tag communication range r is 5 m. In SICP and p-SICP, the reader sets its frame size of the ith request to $f_i = \max\{\hat{n} - m_{i-1}, f_l\}$, where m_{i-1} is the number of IDs that have been collected and $\hat{n}$ is the estimate number of tags that maximizes (4.4). The lower bound f_l, fixed to 50, prevents the frame size from being setting too small or even negative due to the estimation deviation of $\hat{n}$. The initial frame size λ is 50 for the reader. To relieve

---

[1]For the special case when all networked tags are within the coverage of the reader, our protocols naturally become the traditional protocols, literally, because we may actually adopt any existing ALOHA-based RFID identification protocol for collecting IDs within the reader's neighborhood in place of the operations described in Sect. 4.4.4, as long as the serial number is embedded in ACK.

the tags from estimating the numbers of children they have, we let them use a fixed frame size λ with a default value of 4, but we will also vary it from 2 to 10. The length of each tag ID is 96 bits long. The length of each serial number is $\lceil \log_2 N \rceil$ bits long. The length of each tier number is 4 bits long. Following the specification of the EPC global Class-1 Gen-2 standard [1], we set the length of any types of requests to 20 bits, and set ACK and NAK to 16 bits and 8 bits, respectively. In SICP and p-SICP, the ACK will also include a serial number. For each data point in the figures, we repeat the simulation for 100 times and present the average result.

4.6.2 Children Degree and Load Factor

We first examine the balance of the spanning trees built by CICP and SICP (the spanning tree in p-SICP is built in the same way as SICP). It has significant impact on the worst-case energy cost of the tags. A tag with a larger children degree (or a larger load factor) has to collect (or forward) more tag IDs, resulting in additional energy expenditure. Tags that have the largest children degree or load factor may become the energy bottleneck in the network. If the residual on-tag energy is exhausted before the completion of the protocol, the network may even be partitioned due to dead tags.

Figures 4.10 and 4.11 present the maximum children degree and the maximum load factor in the spanning trees built by CICP and SICP, respectively. As the number N of tags in the system becomes larger, the increase in these worst-case numbers under CICP is a lot faster than the increase under SICP, indicating a much balanced tree for the latter. For example, when $N = 10,000$, the maximum children degree and load factor in CICP are 83 and 1969, and those numbers in SICP are only 14 and 165.

Fig. 4.10 Maximum children degree in the spanning tree

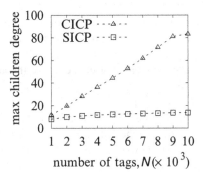

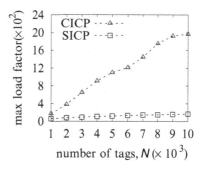

Fig. 4.11 Maximum load factor of tags in the spanning tree

4.6.3 Performance Comparison

We compare the performance of CICP, SICP, and p-SICP in Fig. 4.12, where the first plot shows the protocol execution time in terms of number of slots used, the second plot shows the average number of bits sent per tag, and the third plot shows the average number of bits received per tag. SICP uses a comparable number of slots as CICP, while p-SICP needs a much smaller number of slots than SICP, as we expect. The energy costs of SICP and p-SICP are very close. Both are much smaller than that of CICP, thanks to serialization for collision reduction. For example, when $N = 10,000$, the numbers of bits sent/received per tag in CICP are 8783 and 412,218, whereas those numbers are just 862 and 54,871 for SICP, respectively, which represent 90.2 and 86.7 % reduction over CICP. Because of the request aggregation technique, the average number of bits sent per tag in p-SICP is slightly smaller than that of SICP. But each tag in p-SICP receives slightly more bits on average than SICP. The reason is that a tag in SICP can inform its non-parental neighbors to sleep for a certain duration without receiving the IDs unnecessarily; in contrast, the transmission pipelining of p-SICP requires every neighbor to receive what a tag transmits in each slot. For p-SICP, its numbers of bits sent/received per tag are 634 and 63,716, which represent 92.8 and 84.5 % reduction over CICP, respectively.

Figure 4.13 shows the maximum numbers of bits sent/received by a tag under the three protocols, respectively. As expected, the most energy-consuming tags spend much less energy under SICP and p-SICP than under CICP. For example, when $N = 10,000$, the maximum numbers of bits sent/received by any tag in CICP are 631,412 and 2,367,899, and those numbers in SICP are 38,273 and 159,431—93.9 % and 93.3 % reduction, respectively.

Fig. 4.12 Performance
comparison between CICP
and SICP

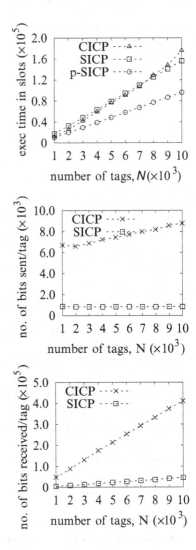

4.6.4 Performance Tradeoff for SICP and p-SICP

Next, we demonstrate a performance tradeoff for SICP and p-SICP controlled by
the value of λ. We set $N = 5000$ and vary λ from 2 to 10. The results are presented
in Fig. 4.14, where the three plots from left to right show the execution time, the
average number of bits sent per tag, and the average number of bits received per tag,
respectively. Similar time-energy tradeoff can observed in both SICP and p-SCIP.
As the value of λ increases, the execution time increases, but the energy cost for
sending and receiving decreases. However, the time increases almost linearly, but
the decrease in energy flattens out, suggesting that a modest value of λ is preferred.

Fig. 4.13 *Upper plot*:
maximum number of bits sent
by any tag in the system.
Bottom plot: maximum
number of bits received by
any tag in the system

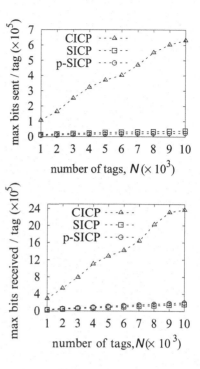

4.6.5 Time-Efficiency Comparison of SCIP and p-SICP

The analysis in Sect. 4.5 demonstrates that the techniques of request aggregation
and transmission pipelining can reduce the time of ID collection, particularly for
tags with large tier numbers. We use simulations to verify this conclusion. We vary
the radius of the circular area, where 5000 networked tags are randomly distributed,
from 50 to 100 m at steps of 10 m. A larger radius of the distribution area means
there are more tiers in the system, resulting in a larger height of the spanning tree.
Figure 4.15 compares the execution time of SICP and p-SICP. We can see that
the gap between the execution time of SICP and p-SICP becomes larger with the
increase of the radius. When the radius is large, p-SICP cuts the execution time of
SICP by more than half.

4.7 Summary

This chapter discusses the tag identification problem in the emerging networked
tag systems. The multihop nature of networked tag systems makes this problem
different from the tag identification problem in RFID systems. Two tag identification
protocols are designed with three important findings. The first finding is that

Fig. 4.14 Execution time
and energy cost of SICP and
p-SICP with respect to λ,
when $N = 5000$

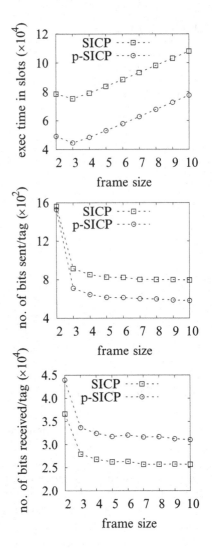

the traditional contention-based protocol design incurs too much energy overhead
in networked tag systems due to excessive collision. The second finding is that
load imbalance causes large worst-case energy cost to the tags. We address these
problems through serialization and probabilistic parent selection based on serial
numbers. The third finding is that the techniques of request aggregation and ID-
transmission pipelining can significantly improve the time efficiency of serialized
ID collection.

Fig. 4.15 Comparison of execution time of SICP and p-SICP when 5000 networked tags are randomly distributed over a circular area with different radiuses

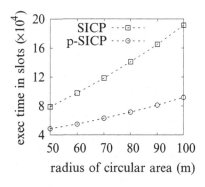

References

1. EPC Radio-Frequency Identity Protocols Class-1 Gen-2 UHF RFID Protocol for Communications at 860MHz-960MHz, EPCglobal (2011). Available at http://www.epcglobalinc.org/uhfclg2
2. Gorlatova, M., Margolies, R., Sarik, J., Stanje, G., Zhu, J., Vigraham, B., Szczodrak, M., Carloni, L., Kinget, P., Kymissis, I., Zussman, G.: Prototyping energy harvesting active networked tags (EnHANTs). In: Proceedings of IEEE INFOCOM Mini-Conference (2013)
3. Han, H., Sheng, B., Tan, C.C., Li, Q., Mao, W., Lu, S.: Counting RFID tags efficiently and anonymously. In: Proceedings of IEEE INFOCOM (2010)
4. Johnson, N.L., Kotz, S.: Urn Models and Their Application: An Approach to Modern Discrete Probability Theory. Wiley, New York (1977)
5. Kinget, P., Kymissis, I., Rubenstein, D., Wang, X., Zussman, G.: Energy harvesting active networked tags (EnHANTs) for ubiquitous object networking. IEEE Trans. Wirel. Commun. **17**(6), 18–25 (2010)
6. Kodialam, M., Nandagopal, T.: Fast and reliable estimation schemes in RFID systems. In: Proceedings of ACM MobiCom (2006)
7. Kong, L., He, L., Gu, Y., Wu, M., He, T.: A parallel identification protocol for RFID systems. In: Proceedings of IEEE INFOCOM, pp. 154–162 (2014)
8. Luo, W., Qiao, Y., Chen, S.: An efficient protocol for RFID multigroup threshold-based classification. In: Proceedings of IEEE INFOCOM, pp. 890–898 (2013)
9. Myung, J., Lee, W.: Adaptive splitting protocols for RFID tag collision arbitration. In: Proceedings of ACM MOBIHOC (2006)
10. Nguyen, C.T., Hayashi, K., Kaneko, M., Popovski, P., Sakai, H.: Probabilistic dynamic framed slotted ALOHA for RFID tag identification. Wirel. Pers. Commun. **71**, 2947–2963 (2013)
11. Onat, I., Miri, A.: A tag count estimation algorithm for dynamic framed ALOHA based RFID MAC protocols. In: Proceedings of IEEE ICC, pp. 1–5 (2011)
12. Qian, C., Liu, Y., Ngan, H., Ni, L.M.: ASAP: scalable identification and counting for contactless RFID systems. In: Proceedings of IEEE ICDCS (2010)
13. Shahzad, M., Liu, A.: Every bit counts - fast and scalable RFID estimation. In: Proceedings of ACM MOBICOM (2012)
14. Shahzad, M., Liu, A.X.: Probabilistic optimal tree hopping for RFID identification. In: Proceedings of ACM SIGMETRICS, pp. 293–304 (2013)
15. Sheng, B., Li, Q., Mao, W.: Efficient continuous scanning in RFID systems. In: Proceedings of IEEE INFOCOM (2010)
16. Zhang, M., Li, T., Chen, S., Li, B.: Using analog network coding to improve the RFID reading throughput. In: Proceedings of IEEE ICDCS (2010)

Printed in the United States
By Bookmasters